CIVILIAN MISSION

ISBN: 979-8-218-31884-0

C.A. Cross & Associates, LLC
PO Box 240014, Honolulu, HI 96824

CIVILIAN MISSION:
The 3-Year Guide for
Military Professionals
Planning Civilian Careers

CHERYL A. CROSS

Foreword by Chase Hughes
US Navy Veteran and CEO of Applied Behavior Research

Publisher: C.A. Cross & Associates, LLC

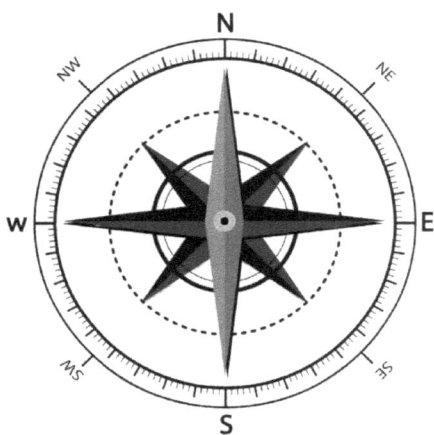

To my father,
Homer "Dillon" Cross
(1928-2022)

U.S. Army Korean War Veteran and Corporate Executive
whose motto,
"Do More, Be Better, and Give Back"
inspired this book.

"America was not built on fear.
America was built on courage,
on imagination,
and an unbeatable determination
to do the job at hand."

Former U.S. President, Harry S. Truman
1949 Inaugural Address

ACKNOWLEGMENTS

I would like to express my heartfelt gratitude to the following individuals who have inspired or contributed to the creation of this 3-year guidebook:

- My late father, Dillon Cross: His two-year draft service during the Korean War reshaped his life, equipping him with the skills for college and his incredible career. His personal Mil2Civ story influenced the creation of this book, his legacy gift helped finance it, and his steady belief in me was the fuel force behind it.

- My son, Steffen Cross-Cohen, for his creative ideas and feedback, and to my sister Karen Cross, M.D. and my uncle, US Army Veteran, Jerry Cathey for their unwavering support and encouragement.

- My dedicated editor and US Air Force Veteran, Mark Antony Rossi, for his timeliness, efforts and understanding of the material. To my friend, colleague and US Army Veteran, Emily Layman-Perkins for the polish.

- US Navy Veteran, Behavior Panel Host, Neuroscience and Behavioral Expert, Chase Hughes for planting the seed for a 3-year timeline and graciously writing the foreword to the book.

- To my State of Hawaii Workforce Development Council colleagues who value the importance of a strong Veteran workforce along with the numerous State agencies and non-profit organizations that help our military workforce transition successfully to civilian careers in Hawaii and throughout the U.S.

- My HR and Recruiting colleagues in the government and civilian spaces who interview and facilitate jobs for the readers of this book each and every day. You are the true warriors who bridge the ever-widening gap between post-military and civilian careers for the 200,000+ transitioning military-to-civilian job applicants each year. Fight on!

Special thanks to my personal mentors, champions, and advocates:

Jon Paul Akeo (Mil2Civ Advocate and HR Leader)
Rex Jordan (Former US Air Force)
Bob Lietzke (Former US Air Force and former Military Spouse)
Michael Collat (Former US Air Force)
Maurice Anstead (Mil2Civ Advocate)
Sean Kelly (Former US Army)
Andrew Gilbert (Former US Marine Corps)
Susanne Goett (Mil2Civ Advocate and TA Leader)
Daniel Brooks (Mil2Civ Advocate and TA Leader)
George Bernloehr (Former US Navy and TA Leader)
Laura Schmiegel (Mil2Civ Advocate, former Military Spouse and TA Leader)
Vanessa Machin-Perez (Former Australian Army, MilSpouse and Mil2Civ Advocate)
Steve Janke (Former US Marine Corps and TA Leader)
Col. Pamela L. Ellison (Hawaii Army National Guard)
Alan Hayashi (Former Hawaii WDC Chair and Mil2Civ Advocate)
Brian Tatsumura (Former Hawaii WDC and EE Committee Chair)
Bennette Misalucha (Former State Senator and Executive Director, Hawaii WDC)
Trang Malone (Former US Army, Military Spouse, Hawaii WDC)
Marisol Maloney (Former US Navy, MilSpouse and TA Leader)
Brandi Brickler (Former US Navy and former Military Spouse)
Ada Barry (Military Spouse)
Elizabeth "Liz" Garcia (Mil2Civ Advocate and Military Spouse)
and the MFSC Transition Team at Joint-Base Pearl Harbor-Hickam (2017 to now)

And to all of the authors, coaches and experts who contributed quotes to this book; may our voices ring louder together as we create important connections for all Mil2Civ career-seeking individuals.

Thank you,

Cheryl A. Cross

Cheryl A. Cross   November 11, 2023

FOREWORD

In 2018, as the weight of retirement bore down on me with just 50 days to go, I found myself clutching a résumé at my kitchen table, pondering a $14.90/hour job at a shipyard.

Two decades in the military had cocooned me in security; a consistent paycheck, a buffer from the whims of the market, and the assurance of employment. Civilian life? What will that even look like?

I...just...need...certainty.

That day, as I grasped that résumé, a call to my mentor reframed my perspective. He flipped a switch in me that I'm hoping I can flip inside you right NOW. He reminded me not of what I lacked, but what I had accumulated over the years.

That pivotal conversation made me realize that my preparation for the civilian world had unknowingly started a decade earlier. Every sacrifice, every night spent in online classes instead of at parties, every certification achieved during service—these weren't mere events, but cornerstones, meticulously laying the foundation for my post-military life.

By the end of December, I retired as an E7. Fast-forward three years, and I was listed among Entrepreneur Magazine's top 40 CEOs in America, earning in a month what the Secretary of the Navy made annually - it was surreal. Weird. I'm not telling you this to brag. I'm telling you this because if I can pivot like that, so can you.

This trajectory wasn't just serendipity—it was guidance from a mentor who shifted my path. I offer this book as a compass for you, hoping to channel that guidance forward. Let these words resonate profoundly, shaping your perspective and actions.

Yes, your military service is a significant chapter in your life, but don't let it be your only one. The starting gun fires now.

The worst insult someone can call me is 'lucky'. There's no luck here. It's only effort. In the words of Thomas Jefferson, "The harder I work, the more luck I seem to have."

Understand this truth: Your transition to civilian life began the moment you put on that uniform. Every endeavor, every accolade, and every habit cultivated will shape the narrative of your post-military journey.

While the prospect of civilian life may seem distant, it approaches a LOT faster than you can imagine. Let your military achievements be a chapter, not the entirety, of your story. Your civilian life has the potential to be magnitudes more significant and rewarding. And the sooner you commit to that vision, the more fruitful it becomes. Seize every opportunity that shapes this approaching chapter.

You swore an oath to defend this nation against all enemies. In civilian life, your most formidable adversary is the reluctance to act towards your future.

Your peers might not grasp why you're tirelessly honing skills or refining your résumé. Extend this book as a bridge, or share the wisdom, so they, too, can transition from military to civilian life, not just seamlessly, but spectacularly.

This book you're holding is a mentor, a guide, and an assistant. The more you decide to use it and take action, the more you will find yourself experiencing the 'luck' that people talk about.

Chase Hughes, CEO
Applied Behavior Research
www.chasehughes.com

Chase Hughes, drawing on two decades of experience in the US Navy, has transformed his military training in human intelligence into a breakthrough career in behavior analysis and neuroscience. He is the author of many impactful books, including his Best-Selling book, "The Ellipsis Manual." Just a few years after his military transition, he was named among the top 40-under-40 CEOs in America. Today, he continues to train government agencies and companies, write more books, guest on popular news, and talk shows, including Dr. Phil and create weekly videos with his behavioral expert colleagues, Greg Hartley (Fmr. US Army), Scott Rouse and Mark Bowden for The Behavior Panel on YouTube. Chase Hughes is a testament to leveraging military experience in impactful ways, creating a dynamic and successful post-military career.

TABLE OF CONTENTS

INTRODUCTION
My Story: From DOD Corporate Recruiter to Your Guide

My name is Cheryl Cross, and for the next three years, I'll be your trusted guide. Now, you might be wondering why a civilian businesswoman like me is writing a book on Mil2Civ's career transition. Well, I've had a front-row seat to hiring former military in corporate America, and I can tell you one thing for sure – there's a better way, and I'm here to show you how.

Like your career ahead, mine has taken exciting turns. Starting in broadcasting, working as an on-air radio talent and later music director in my hometown of Las Vegas, Nevada, I was transferred by Geffen Records to work in emerging markets as a radio promoter in Los Angeles, California. From there, I expanded to marketing and media relations for a Warner/Electra/Atlantic-based distributor. After a few years, and with the explosion of the internet, I dove into emerging tech and the world of competitive intelligence analysis. Wanting to grow into new industries, a friend suggested I share best-practices and help streamline the research division for a recruiting firm. It was there that I developed a passion for recruiting and making connections between people, places, and careers.

After 15 years of independent consulting in communications, marketing, research, and recruiting, I joined Booz Allen Hamilton, a Global Fortune 500 technology consulting firm that has served the Department of Defense since 1940.

My move to Hawaii to support the firm's clients in the Indo-Pacific AOR gave me a front seat to the issues surrounding military to civilian career transition. As a senior corporate recruiting specialist, I was invited to participate in countless career fairs, job panels, TAPS/ETAPS classes and keynote presentations. In my role, I interviewed hundreds (if not thousands) of you who were looking to carve out a corporate career, post-military. I had the privilege and honor to hear your questions, field your concerns, and saw firsthand the numerous challenges and frustrations you faced.

As I delved deeper into the experiences of transitioning service members, I noticed some common themes. Many individuals faced challenges due to the lack of a structured career plan, which often resulted in uncertainty and missed opportunities. Additionally, I observed a widespread lack of salary negotiation skills for soon-to-be Veterans, especially women, minorities, and people of color, leading to potential undervaluation in the job market. There was also limited understanding of industry or market research, which hindered informed decision-making during the transition process. These recurring themes highlighted the critical areas where a pre-planning approach would serve to ensure a successful transition to civilian careers.

This three-year pre-planning guide gives you ample time to develop a strategic and tactical path to your next career. You will have details on what you can do and learn each year, to gain a stronger sense of self and where you want to go. Also, you will hone negotiation skills, and gain insights into different industries, ensuring a more seamless and successful transition to civilian life.

Don't wait until the last year of military service; start your journey to a thriving civilian career now.

PREFACE

The transition from a military to a civilian career is more than a change in jobs; it's a complete lifestyle shift. As soldiers, sailors, Airmen, and Marines, you've been part of a tightly knit community, shared common experiences, and lived by a distinct set of values and rules. Now, you're preparing to venture into a new and unfamiliar territory where the terrain and culture are significantly different. It's a critical journey, one that requires planning, strategy, and a clear understanding of what to expect.

Why This Transition Matters

For many, leaving the military isn't a matter of simply taking off a uniform and putting on corporate garb or a business suit. It's about learning to navigate the often-murky waters of job applications, resumes, and interviews, understanding corporate culture, and translating your military skills into terms that civilian employers can appreciate and comprehend. This transition, with all its challenges, is of paramount importance for numerous reasons.

#1: There is an economic imperative.

Having committed years to military service, you've acquired a unique set of skills and a perspective that can be an asset to many employers. However, effectively translating those into civilian job roles isn't always straightforward. This transition, when executed well, ensures economic stability and job satisfaction for veterans.

#2: There's the psychological aspect.

Leaving the military can be a momentous change, affecting your identity, relationships, and daily routines. A successful transition will look different for everyone and is very individual, but the goal of this book is to help you gain the confidence you need for the journey ahead. These pages are filled with tips and strategies for you to maintain a positive self-image, create a strong sense of purpose, and foster personal growth in your post-military life.

Understanding the Challenges Ahead
This career transition presents a unique set of challenges, ranging from tangible, like crafting an effective resume, to more abstract, such as adjusting to different work cultures and managing emotions associated with leaving the military.

A significant challenge is effectively translating your military skills and experience into civilian terms. You might have been an 'infantry squad leader' in the military, but how does that translate to a job role a civilian HR or hiring manager would understand?

Another important, but sometime overlooked challenge lies in understanding and adapting to civilian work culture. Military and civilian workplaces often operate under different sets of norms, values, and expectations. In the military, rank, and chain of command dictate many interactions, whereas civilian workplaces tend to be more collaborative with less formal hierarchies. In short, you are moving from a very predictive career environment into a very flexible and organic environment.

Preparation is Key
If there's one thing your military experience has instilled in you, it's the value of preparation. Just as you wouldn't go into a mission without a clear plan, you shouldn't approach your career transition without one either. That's why this guide is designed to be used starting at least three years before your military separation or retirement.

This book is aimed at guiding you through this process step-by-step, equipping you with the necessary tools and knowledge to navigate this new journey. It's about leveraging your existing skills, understanding the landscape you're stepping into, and preparing you mentally and emotionally for the changes ahead.

Over the next few chapters, we will dive deeper into these topics, offering practical advice and strategies. This journey might be challenging but remember: every new beginning comes from some other beginning's end.

Here's to your new beginning.

The Role of This Guide: How to Use This Book

This guide is more than just a book; it's a roadmap, a tool designed to help you navigate the path from military service to civilian life. It is truly your "Civilian Mission," and its purpose is twofold: to enlighten you about the nature of the civilian job market and to equip you with practical skills and strategies for a successful transition.

To get the most out of this book, approach it as a journey rather than a destination. Start reading and implementing the strategies as early as three years before your projected separation or retirement date. Each chapter has been systematically arranged to correspond with specific milestones along your transition timeline.

The initial chapters are focused on laying the groundwork and understanding the civilian job market landscape. They will help you identify your transferable skills, initiate your networking efforts, and start considering potential career paths. As you progress, the content becomes more tactical, addressing job applications, resumes, interviews, and the ins-and-outs of civilian work culture.

Throughout the book, you'll discover resources and links tailored for you at specific stages of your transition from the military into potential civilian roles, coupled with contemplative exercises in each chapter. These exercises are designed to offer you an introspective journey, allowing you to explore and align your military experiences with your civilian aspirations. The resources provided have been carefully curated to present tools, strategies, and best practices that will aid in your transition journey. By incorporating these exercises and tools, the book not only offers knowledge but also encourages personal growth and understanding. This book was written with the hopes that it will serve as guide for at least three years or even longer, ensuring the insights support you through the initial Mil2Civ career transition and as you navigate the broader and longer journey of your civilian career.

Finally, remember this guide is not meant to be consumed in one go. Take your time. Reflect on the advice given. Implement the strategies step by step. And remember, this guide is here to help you, but it's your commitment and proactive planning that will ultimately determine your success in transitioning from military service to a rewarding civilian career. Your journey from military service to a fulfilling civilian career is a unique "Civilian Mission," and this book is dedicated to ensuring you navigate it with confidence and clarity.

CHAPTER 1:
THE JOURNEY AHEAD -
WHAT TO EXPECT

1.1 Understanding the Civilian Job Landscape

Starting your transition from military service to civilian life three years in advance, instead of the suggested one year, puts you at a significant advantage. This early start allows you to take a deliberate and strategic approach to understanding and preparing for the civilian job market. This is not a race, but a journey that requires careful planning and thoughtful execution. So, let's take your first step in understanding the terrain of the civilian job market.

The civilian job market is an intricate matrix of industries, sectors, and professions. It's quite different from the structured path of military career progression. This diversity may seem overwhelming initially, but it's also an invitation to research the limitless possibilities, opportunities and creative ventures that align with your skills, interests, and aspirations.

It's crucial to start familiarizing yourself with various sectors such as Technology, Healthcare, Finance, Energy, Government and or, Defense, Education, Manufacturing or other emerging sub-sectors like Artificial Intelligence, Renewable Energy, Biotechnology, E-Commerce and more. Each industry and sub industry has its unique features, opportunities, trends, and challenges. Remember, you have a three-year runway to explore these sectors, allowing you to understand where your current skills and experiences could fit and how you might start training towards future goals in emerging markets.

Wherever you start your research, you'll encounter a multitude of job roles. Many might seem unfamiliar at first but don't be quick to dismiss them. The skillsets you acquired during your military service, like leadership, discipline, teamwork, crisis management, and many others, are highly valued across all sectors. You have time to learn how these skills translate into civilian roles, so use them wisely.

Staying updated with current economic factors and industry trends is essential. Over the next three years, certain industries will grow and create new sub-industries while others may retract or become obsolete. Some will be influenced by technological advancements, policy changes, or shifts in market needs. Use this time to spot trends, identify growth sectors, and understand how your skills can fit into the future job market, whatever that may look like to you today.

Understand that the civilian job market is competitive. But don't let this intimidate you. Your three-year lead time means you are gaining the confidence to not only know what you want but ask for your worth. Your strategy today will prepare you to succeed, far better than most candidates. The military has equipped you with unique experiences and a skill set that can make you stand out. Your leadership roles in high-pressure situations, your ability to adapt to rapidly changing environments, and your strong work ethic are qualities that employers highly value. Your experience with cutting-edge technology, logistics, teamwork, and effective communication sets you apart in various industries. These are just some of the assets you bring to the table as you embark on your civilian career journey. Your mission over the next few years is to learn how to articulate these skills and experiences effectively to potential employers, a topic we will delve into in later chapters.

Remember, embarking on this journey three years in advance is your strength. The civilian job market may feel like a new frontier, but with time, research, and strategic planning, you'll transition from unfamiliarity to mastery with great confidence. There's a wealth of opportunities waiting for you, and the time to seize them starts today. Working in advance is the best way to prepare for your future and is the first step on your journey to a fulfilling civilian career. Get ready to explore, learn, and prepare for the road ahead.

1.2 Common Challenges for Transitioning Personnel

As you embark on your transition journey, it's essential to be aware of some of the common challenges that transitioning personnel often face. Understanding these hurdles ahead of time not only helps you prepare effectively but also places you in a position of strength as you can plan to tackle these issues well in advance. You have a significant advantage with a three-year start, providing ample time to strategize and implement solutions to these challenges.

One of the most common difficulties veterans encounter is translating military skills and experiences into civilian language. Military jargon and acronyms might not make sense to a civilian employer. Over the next three years, you can take the time to learn how to effectively communicate your skills and experiences in terms civilian employers understand, ensuring your unique qualifications are not lost in translation.

Another challenge is understanding and adapting to the civilian workplace culture. Civilian work environments can be a far cry from military settings, with less formal hierarchies and different social norms. Your three-year runway gives you a valuable opportunity to interact with civilians in professional settings, understand their work culture, and adjust your expectations.

Job searching itself can pose a significant challenge. Unlike the military, where career progression is relatively structured, the civilian job market requires proactive job hunting, networking, and keeping an eye on industry trends. Starting three years out allows you to gradually build your job-hunting skills, grow your network, and understand the landscape without feeling rushed or pressured.

Another issue military personnel often grapple with is determining their worth in the civilian job market. Salary negotiations are often unfamiliar territory for many soon-to-be or current veterans. However, you can utilize your three-year head start to research salaries for positions that align with your skills, experience, and industry preferences. During this time, you'll also research locations and how the local cost of living in individual cities, both in CONUS and in OCONUS countries, and how specific locations may align to your career goals. Using the steps and exercises in this guide, you will fine-tune your negotiation skills and gain strength and practice for those critical conversations.

Finally, many veterans face emotional and psychological challenges during their transition. Leaving the military can be a significant shift in identity, structure, and community. You can use these three years to prepare mentally and emotionally for this change, seeking support as needed from mental health professionals, peers, and veteran support organizations.

Remember, while these challenges are common, they are far from insurmountable, especially with your three-year head start. This period is a unique advantage, a buffer that allows you to face these challenges head-on and emerge victoriously. The journey might not always be easy, but with preparation, resilience, and the right support, you're well-equipped to navigate the path to your successful career transition.

1.3 The Importance of Starting Early

The decision to start your career transition three years in advance is not just a choice; it's a tactical move that gives you an edge. It offers ample time to plan, prepare, and execute your transition effectively, ensuring you have the best chances of success. Let's break down the importance of this early start with some tangible bullet points and facts. I've added both facts and my own corporate recruiter observations to underscore the importance of these key areas.

Career Exploration: With three years, you have the luxury of time to explore various industries, job roles, and career paths, helping you find the perfect fit for your skills and aspirations.

- Fact: Research shows that career explorers are more likely to be satisfied and successful in their careers due to the increased self-awareness and clarity they gain during their exploration.
- Recruiter observation: Candidates who know who they are and what they want are far more confident, and successful in the interview and hiring process.

Skills Translation and Acquisition: A three-year head start allows ample time to translate your military skills into civilian terms and to acquire any additional skills or certifications required for your chosen career path.

- Fact: According to a study published in 2016 from a survey by the Center for a New American Security (CNAS) a non-partisan research institution in Washington, DC reported 60% of employers report that veterans do not effectively translate their military skills for the civilian job market.
- Recruiter observation: Not all recruiters have DOD experience and may have a hard time understanding the value your career brings to their organization. While many are coaches and advocates, most are looking for keywords that

align with their job requisitions. Be prepared to translate in both your resume and your interviews.

Networking: Starting early allows you to build and expand your professional network, creating relationships that could potentially lead to job opportunities down the line.
- Fact: According to LinkedIn, 85% of jobs are filled through networking. This underscores the importance of starting networking efforts early.
- Recruiter observation: Knowing the stages of networking and what your 'ask' will be at the different phases of your three-year plan is essential. You will learn tips and tricks in the following chapters about the types of networking and how to build a robust professional circle of 'trust.'

Understanding Civilian Work Culture: Three years provides a significant timeframe to understand and adapt to civilian work culture, easing the cultural transition and preparing you to start your new job.
- Fact: A 2015 study by the D'Aniello Institute for Veterans and Military Families at Syracuse University found that cultural differences between military and civilian life were among the most significant challenges faced by transitioning veterans.
- Recruiter observation: Much like an immersive language program, working in a corporate culture as a civilian will give you a best-in-class opportunity to learn, but it's helpful to start researching the differences between military and corporate culture in advance of your start date.

Job Searching and Application Preparation: Early preparation enables you to craft compelling resumes, cover letters, and a LinkedIn profile. You will also learn effective job searching techniques, and practice interviewing and salary negotiation skills.

- Fact: The average job search takes about five months, according to a study by the U.S. Bureau of Labor Statistics. Starting early ensures you're ready when the time comes.
- Recruiter observation: Preparing lends itself to confidence-building, and nothing is more impressive than a confident candidate interviewing for a job. I always appreciated those who did their research on what they wanted, where they wanted to live and the skills they brought to the table (in civilian speak), making for an excellent interview process.

Emotional and Psychological Preparation: Transitioning from military to civilian life can have emotional and psychological implications. Three years provide a buffer for you to mentally prepare for this life-altering change and seek support as needed.

- Fact: The Substance Abuse and Mental Health Services Administration reports that veterans often experience challenges such as depression, anxiety, and PTSD, highlighting the importance of emotional preparation and support during the transition.
- Recruiter observation: As a civilian working in both government contract and civilian corporate and nonprofit environments, I may not fully comprehend the emotional and psychological journey of leaving the military and entering a brand-new environment. However, I've observed that those who take the time to prepare emotionally tend to adapt more smoothly and find greater success in their civilian careers.

In essence, a three-year head start isn't just a nice-to-have—it's a game-changer. It puts you in a position of strength and ensures that when the day of transition arrives, you're not just ready, but you're poised for success. So, as you embark on this journey, remember that every step you take now, no matter how small, is a step towards a successful civilian career.

CHAPTER 2:
LAYING THE GROUNDWORK -
3 YEARS OUT

2.1 Building Self-Awareness: Identifying Transferable Skills and Interests

As you prepare to transition from the military to civilian life, self-awareness is a critical first step. This involves identifying your transferable skills and understanding your interests. This is not only essential for finding a career that aligns with your capabilities but also crucial in identifying a path that will bring you satisfaction and fulfillment.

Your military experience has equipped you with a multitude of skills, regardless of your rank. However, it's essential to translate these skills into language and concepts that civilian employers understand.

Whether you've served as an E-4 Specialist or Corporal, Petty Officer Third Class, Senior Airman or an E-7 Sergeant First Class, Chief Petty Officer, Master Sergeant, or Gunnery Sergeant you've not only gained significant technical skills but gained valuable expertise in areas such as individual and group leadership, team coordination and planning, problem-solving, flexibility and adaptability.

Similarly, if you've been commissioned as an O-4 Major or Lieutenant Commander, or an O-6 Captain or Colonel, you've also acquired deep technical expertise while demonstrated exceptional leadership, strategic thinking, decision-making, along with budgeting and project management skills. These competencies are highly transferable and sought after in various civilian career paths and industries.

To identify your transferable skills, let's work together to consider your military experiences and duties. What responsibilities did you have? What tasks did you perform regularly? What skills were required for these tasks? Make a list of these skills and consider how they can apply to a civilian work context.

In addition to skills, it's crucial to identify your interests. Think about what you enjoy doing and what you're passionate about. This could be anything from problem-solving and analytical tasks to mentoring and training others, or maybe something creative or technical. Your interests can help guide you toward a career that you'll find fulfilling.

To help you through this process, let's work through the following exercise:

Identifying Transferable Skills and Interests

Military Experiences and Responsibilities:
1. List your roles, duties, and responsibilities during your military service.
2. Detail the skills required to fulfill these responsibilities.

Examples:

E-4 Corporal or Specialist (Enlisted):
Roles and Duties: As an E-4, I served as a team leader responsible for supervising and mentoring junior soldiers. I also conducted training and ensured the readiness of my team.
Skills: Leadership, team management, training coordination.

E-7 Sergeant First Class (Enlisted):
Roles and Duties: In my role as an E-7, I held a senior leadership position, overseeing multiple teams, and was responsible for mission planning and execution.
Skills: Strategic planning, leadership under pressure, decision-making.

O-4 Major (Officer):
Roles and Duties: As an O-4, I served as a company commander responsible for the overall mission success of my unit. I also managed personnel, resources, and budgets.
Skills: Command leadership, resource management, strategic thinking.

O-6 Colonel (Officer):
Roles and Duties: In my position as an O-6, I held a high-ranking leadership role, overseeing a large organization, and advising senior leadership on strategic matters.
Skills: Executive leadership, strategic planning, policy development.

Transferable Skills and Applications and Interests:
1. From your list above, identify skills that can apply to civilian roles.
2. Write it in first-person as if you are telling a story.

Examples:
E-4 Corporal or Specialist (Enlisted):
Discipline: In my role as an E-4, I learned the importance of discipline through adhering to military protocols, which can be applied to any job requiring strong work ethics and reliability.

Problem-Solving: As a team leader, I often encountered unexpected challenges during training exercises and had to quickly adapt and find solutions. This skill is valuable in roles that require critical thinking and problem-solving.

E-7 Sergeant First Class (Enlisted):
Communication: In my leadership role as an E-7, effective communication was crucial for conveying mission objectives and ensuring the team's understanding. This skill is applicable to any job involving team coordination and clear communication.

Conflict Resolution: I gained experience in resolving conflicts within my team, which can be transferred to roles that require conflict resolution and mediation.

O-4 Major (Officer):

Project Management: As an O-4, I oversaw complex missions, which involved project planning, execution, and evaluation. This skill can be useful in civilian project management roles.

Data Analysis: I often had to analyze data and intelligence to make informed decisions. This analytical skill is applicable in various industries, particularly those involving data-driven decision-making.

O-6 Colonel (Officer):

Strategic Thinking: In my position as an O-6, I developed a strong ability for long-term strategic thinking. This skill is valuable for senior management and executive roles in organizations where strategic planning is essential.

Leadership Development: I mentored and developed junior officers and leaders. This experience can be translated into roles involving leadership development and training in civilian organizations.

Create more skills or expand on each skill by writing down more examples of how you've applied it during your service. Continue writing in the first person so these points can be easily adapted into your elevator pitches or future interviews. NOTE: These points will be very helpful if you are faced with 'behavioral questions' during the interview. These sometimes start with, "Tell us a time where you…" The essence of these questions is to drill down three points: How did you behave, what were the skills you deployed, and what was the solution or outcome?

Examples:

E-4 Corporal or Specialist (Enlisted):

Skill: *Discipline*

Example 1: I consistently followed the military code of conduct, ensuring I was always punctual, well-groomed, and prepared for duties.

Example 2: During training exercises, I maintained strict adherence to safety protocols, preventing accidents and ensuring the well-being of my team.

Skill: *Problem-Solving*

Example 1: In a field exercise, our team faced logistical challenges. I devised an alternative route that allowed us to complete the mission successfully.

Example 2: During a technical malfunction, I quickly identified the issue with our equipment and implemented a temporary fix, preventing mission delays.

E-7 Sergeant First Class (Enlisted):

Skill: *Communication*

Example 1: I regularly conducted briefings to convey mission objectives, ensuring that all team members understood their roles and responsibilities.

Example 2: During joint exercises with international forces, I facilitated effective communication among diverse groups, fostering cooperation.

Skill: *Conflict Resolution*

Example 1: In a high-stress environment, I mediated a dispute between team members, addressing concerns and maintaining team cohesion.

Example 2: During a joint operation, I collaborated with counterparts from different branches, diplomatically resolving differences in approach and strategy.

O-4 Major (Officer):
Skill: *Project Management*
Example 1: I led a complex training exercise, overseeing logistics, personnel, and timelines to ensure a seamless execution.
Example 2: While deployed, I managed construction projects critical to the mission, coordinating resources and contractors to meet tight deadlines.

Skill: *Data Analysis*
Example 1: I analyzed intelligence reports to identify patterns and trends, enabling informed decision-making for mission planning.
Example 2: I utilized data analytics tools to assess the effectiveness of training programs, optimizing resource allocation for maximum impact.

O-6 Colonel (Officer):
Skill: *Strategic Thinking*
Example 1: As a senior leader, I formulated long-term strategies for force readiness, adapting to evolving threats and geopolitical shifts.
Example 2: I guided the development of a comprehensive cybersecurity strategy to safeguard critical military assets from cyber threats.

Skill: *Leadership Development*
Example 1: I mentored and coached junior officers, assisting in their professional growth and preparing them for leadership roles.
Example 2: I designed and implemented leadership training programs, focusing on cultivating leadership qualities in emerging leaders.

Interests:

Staying in the first person (I, my), list the aspects of your work that you most enjoyed during your service.

These could be specific tasks, types of projects, or general aspects like teamwork, leadership, problem-solving, etc. While the examples below only list one or two interests, you can write several and follow with an "I" or first-person statement to explain where you used it and to what extent.

Examples:

E-4 Corporal or Specialist (Enlisted):

Interest: *Leadership*

Example: I found great satisfaction in leading my squad during field exercises, guiding them through challenging scenarios and ensuring mission success.

Interest: *Teamwork*

Example: Collaborating with my fellow soldiers in high-stress situations was a highlight of my service. We relied on each other, and the sense of camaraderie was rewarding.

E-7 Sergeant First Class (Enlisted):

Interest: *Mentorship*

Example: One of my passions was mentoring junior enlisted soldiers, helping them develop their skills and watching them grow into proficient team members.

Interest: *Strategic Planning*

Example: I enjoyed contributing to the strategic planning process, where I could use my experience to shape the direction of our missions and exercises.

O-4 Major (Officer):

Interest: *Problem-Solving*

Example: I thrived on tackling complex challenges, whether in the field or in logistics. Finding innovative solutions was both fulfilling and intellectually stimulating.

Interest: *Leadership Development*

Example: Watching the officers that I mentored who progressed in their careers to take on leadership roles was deeply rewarding. I had a genuine interest in their growth.

O-6 Colonel (Officer):
Interest: *Strategic Thinking*
Example: I had a strong passion for strategic planning and enjoyed devising long-term strategies to address evolving threats and achieve mission objectives.
Interest: *Policy Development*
Example: Contributing to the development of military policies that had a positive impact on our operations was a particular interest of mine. I appreciated the opportunity to shape our approach.

Now, reflect on how these interests might translate into a civilian career. What kind of roles or industries might align with these interests? Continue writing in the first person using "I" statements as if you are speaking with a corporate recruiter.

BONUS: Add a few civilian roles or industries you think (at this time) may align with these interests.

Examples:
E-4 Corporal or Specialist (Enlisted):
Interest: *Leadership*
Civilian Language: *"I have a strong background in leadership roles, where I've successfully guided teams to achieve their goals, even in high-pressure situations."*

Possible Civilian Roles/Industries:
    a.   Team Leader,
    b.   Project Manager,
    c.   Leadership Development Programs
Potential Industries:
    a.   Manufacturing,

b. Logistics,
c. Healthcare

Interest: *Teamwork*

Civilian Language: *"I excel in collaborative environments, having worked closely with diverse teams to ensure we collectively meet objectives."*

Possible Civilian Roles/Industries:
a. Team Collaboration,
b. Cross-Functional Project Teams,
c. Sales Team Member

Potential Industries:
a. Information Technology,
b. Retail,
c. Marketing

E-7 Sergeant First Class (Enlisted):

Interest: *Mentorship*

Civilian Language: *"I have a passion for mentoring and developing talent, which I've successfully done to enhance team performance."*

Possible Civilian Roles/Industries:
a. Talent Development,
b. Human Resources,
c. Training and Development

Potential Industries:
a. Education,
b. Aerospace,
c. Finance

Interest: *Strategic Planning*

Civilian Language: *"I'm experienced in strategic planning, where I've contributed to the development of effective long-term strategies to achieve organizational goals."*

Possible Civilian Roles/Industries:
    a. Strategic Planning,
    b. Operations Management,
Business Strategy
    a. Potential Industries:
    b. Consulting,
    c. Energy,
    d. Nonprofit

O-4 Major (Officer):

Interest: *Problem-Solving*

Civilian Language: *"I thrive on solving complex challenges and have a track record of finding innovative solutions, which can be applied to enhance business processes."*

Possible Civilian Roles/Industries:
    a. Consulting,
    b. Project Management,
    c. Research and Development
Potential Industries:
    a. Technology,
    b. Engineering,
    c. Healthcare

Interest: *Leadership Development*

Civilian Language: *"I have a genuine interest in nurturing leadership skills in others, as seen through my mentoring and leadership development initiatives."*

Possible Civilian Roles/Industries:
    a. Leadership Development Programs,
    b. Talent Management,
    c. Executive Coaching
Potential Industries:
    a. Education,
    b. Government,
    c. Pharmaceuticals

O-6 Colonel (Officer):

Interest: *Strategic Thinking*

Civilian Language: *"I possess strong strategic thinking abilities, having crafted and executed long-term strategies to address complex challenges and achieve objectives."*

Possible Civilian Roles/Industries:

    a. Executive Leadership,
    b. Corporate Strategy,
    c. Management Consulting

Potential Industries:

    a. Finance,
    b. Defense,
    c. Manufacturing

Interest: *Policy Development*

Civilian Language: *"I've played a key role in shaping policies that positively impact operations, showcasing my ability to influence policy development."*

Possible Civilian Roles/Industries:

    a. Policy Analysis,
    b. Government Relations,
    c. Regulatory Affairs

Potential Industries:

    a. Healthcare,
    b. Energy,
    c. Legal

It's important to underscore the value of these exercises. Writing down your experiences, skills, interests, and career aspirations in the first person is more than just an exercise – it's the first step in advocating for yourself in the civilian job market. By getting into the practice of dissecting your skills and translating them into language that anyone can understand, you're not just enhancing your resume or beefing up your interview chops; you're equipping yourself with a powerful tool to articulate your worth both within and beyond the military.

Remember, these written notes aren't set in stone. You can keep them, add new skills and interests as they come to you, or even start fresh each year, reflecting the evolving landscape of your career goals and the knowledge you'll gain during your three-year career transition journey.

So, embrace this practice, make it a habit, and watch as it empowers you to confidently pursue the career you desire.

2.2 Initial Career Path Considerations including Tech, Artificial Intelligence (AI) Data Science and the Cloud

Assuming you are starting your three-year transition journey in Year One, a crucial step involves exploring possible career paths. An area of significant growth and opportunity lies within technical careers, even for those who don't see themselves as "techy."

The technology sector is expansive, with roles that range from purely technical positions to those that support or intersect with technology, like project management, sales, customer support, or data analysis. What's exciting about this industry is that it is increasingly welcoming to people from non-technical backgrounds, valuing diverse perspectives and problem-solving skills.

Don't be put off by the 'technical' tag. Many roles in this sector need less technical expertise and more analytical thinking, problem-solving abilities, communication skills, and an understanding of users or markets. These are all competencies that your military career has likely equipped you with.

To position yourself optimally for these roles, there are certifications and courses that you can consider. Here are a few examples:

1. **Project Management Professional (PMP):** This globally recognized certification is ideal for those interested in project management roles in the tech industry. It shows that you know how to manage projects efficiently and effectively.

2. **Certified Scrum Master (CSM):** This certification is for those interested in agile project management, particularly in software development environments.

3. **CompTIA A+:** This certification is an entry-level credential for those seeking IT support roles. It covers mobile devices, networking technology, hardware, virtualization, and cloud computing. CompTIA also offers courses and

21

certifications in networking, cyber security, and advanced programming.

4. **Microsoft Certified: Azure Fundamentals:** If you're interested in cloud technologies, this certification can provide a foundational understanding of Microsoft's popular Azure platform.

5. **Google Analytics Individual Qualification (GAIQ):** For those interested in roles that involve website performance analysis or digital marketing, this certification provides a robust understanding of Google Analytics.

Remember, earning these certifications takes time, making your three-year head start all the more advantageous. By identifying your interests and starting the necessary learning now, you can build up an impressive resume that stands out to employers.

As you explore these career paths, maintain an open mind. Reflect on the skills and interests you've identified through the exercises you completed earlier in this chapter. Consider the type of work environment that aligns with your strengths and passions. Keep in mind that you're not merely transitioning to a civilian job; you're embarking on a journey to build a new career that can provide you with satisfaction, fulfillment, and growth. This phase of your transition is all about discovery and education. Make the most of this period to establish a solid groundwork for your future career.

Exploring Artificial Intelligence (AI)

As the digital landscape continues to evolve, AI is becoming a significant player across industries, including healthcare, finance, transportation, and of course, technology. According to the World Economic Forum, Artificial Intelligence (AI) and Machine Learning (ML) will create 133 million new jobs by 2025. It's not just for tech geniuses and software engineers; there are numerous AI roles that people from non-technical backgrounds can excel in.

Roles like AI Project Managers, AI Ethicists, or Data Analysts are examples of positions where your military background—problem-solving, strategic thinking, and leadership—can be highly valuable.

To start learning about AI, you could consider the following entry-level certifications and courses:

1. **IBM AI Foundations for Business:** This course is designed for those who want to learn the basics of AI and its business applications. It's a great starting point for understanding AI without diving into the technical details.

2. **Microsoft Certified: Azure AI Fundamentals:** This certification covers common AI workloads and machine learning concepts, providing a foundation for further learning in AI.

3. **Professional Certificate in Applied AI from IBM:** This professional certificate introduces you to AI and machine learning, helping you understand AI concepts and workflows, and how to build AI models.

If you are seeking a deeper dive into this world, are more experienced in technology, or hold an engineering, computer science or advanced mathematics degree consider these:

1. **Google Cloud Certified: Professional Cloud AI Engineer:** This certification demonstrates your ability to design, implement, and manage AI solutions on Google Cloud Platform.

2. **AWS Certified Machine Learning – Specialty:** This certification validates your ability to design, implement, and manage machine learning solutions on Amazon Web Services (AWS).

3. **Certified Artificial Intelligence Professional (CAIP):** This certification is offered by the Association for

Computing Machinery (ACM) and is designed to assess your knowledge of AI fundamentals and best practices.

4. **Certified Ethical Artificial Intelligence Professional (CEAIP)**: This certification is offered by the Artificial Intelligence Society (AIS) and is designed to assess your knowledge of AI ethics and responsible AI development.

5. **Certified Artificial Intelligence Practitioner (CAIP):** This certification is offered by the CertNexus organization and is designed to assess your knowledge of AI fundamentals and practical AI skills.

In addition to these general AI certifications, there are also many certifications that focus on specific AI technologies or applications. For example, you could earn a certification in machine learning, natural language processing, computer vision, or robotics.

Starting your learning journey into AI now can open up exciting career opportunities in the future. As you progress on this path, remember that AI isn't about replacing human skills but augmenting them. Your military background combined with your knowledge of AI could make you a unique and sought-after candidate in the job market.

Data Analysis and Data Science: Valuable Skills for Military-to-Civilian Career Candidates

One of the ways you can make your resume stand out to potential employers is to showcase any/all data analysis or data science skills. You may not realize that the work you did in active duty as an intelligence operative to a data analysis or science role, or that the cyber or logistics MOS had you crunching data on the regular. That's ok. Let's take a first look at what these two are and see how you may use data science or data analysis on a resume or in an interview.

Data analysis and data science are two related fields, but there are some key differences between the two.

❖ **Data analysis** is the process of collecting, cleaning, and examining data to extract meaningful insights. Data analysts use a variety of tools and techniques to analyze data, including statistical analysis, data visualization, and machine learning.

❖ **Data science** is a broader field that encompasses data analysis, as well as other disciplines such as machine learning, artificial intelligence, and computer science. Data scientists use their skills to build models and algorithms that can be used to solve complex problems and make predictions.

In short, data analysis is about understanding what has happened, while data science is about predicting what will happen.

How is Data Analysis and Data Science Used in the Military?

Enlisted

• Intelligence analysts: Enlisted intelligence analysts use data science and data analysis to collect, clean, and analyze data to identify patterns and trends that can be used to inform military decision-making.

• Cybersecurity analysts: Enlisted cybersecurity analysts use data science and data analysis to identify and mitigate cyber threats.

• Logisticians: Enlisted logisticians use data science and data analysis to manage the movement of supplies and personnel.

• Maintenance technicians: Enlisted maintenance technicians use data science and data analysis to predict and prevent equipment failures.

Officer

- Military intelligence officers: Military intelligence officers use data science and data analysis to develop and implement intelligence plans.

- Cybersecurity officers: Cybersecurity officers use data science and data analysis to protect military networks and systems from cyber-attacks.

- Logisticians: Logistics officers use data science and data analysis to optimize the supply chain and ensure that troops have the resources they need.

- Maintenance officers: Maintenance officers use data science and data analysis to develop and implement predictive maintenance programs.

In addition to these specific examples, data science and data analysis are used in many other ways throughout the military.

You may have used data in any or all of these scenarios: To develop new weapons and technologies, to train personnel, to plan and execute military operations and to safeguard equipment by using predictive measures.

As we established earlier in the Skills and Interests exercise, you have transferable skills that are valuable in data science roles. For example, if your skills listed problem-solving, you would be a good candidate for entry level data analysis.

How to Gain Data Science Skills

There are many ways to gain data science skills. You can take online courses, attend bootcamps, or self-learn using books and online resources.

Here are a few tips for gaining data science skills:

1. Start by learning the fundamentals of data science, such as statistics, machine learning, and data visualization.

2. Gain experience with data science tools and technologies, such as Python, R, and SQL.

3. Build a portfolio of data science projects to demonstrate your skills to potential employers.

If you are starting with no data science/analysis experience, here are a few entry-level certifications that you can consider:

1. **Google Cloud Certified: Data Fundamentals:** This certification provides a foundational understanding of Google Cloud Platform and its data services.

2. **AWS Certified Cloud Practitioner:** This certification provides a foundational understanding of Amazon Web Services (AWS) and its cloud computing services.

3. **Microsoft Certified: Azure Data Fundamentals:** This certification provides a foundational understanding of Microsoft's Azure platform and its data services.

4. **CompTIA Data+:** This certification covers the fundamentals of data analysis, including data collection, cleaning, and visualization.

5. **IBM Professional Certificate in Data Science:** This certificate program provides a comprehensive introduction to data science, including data analysis, machine learning, and data visualization.

Advanced Data Science Certifications

If you have some experience in data science and analysis, you may want to consider pursuing more advanced certifications. Here are a few examples:

1. **Google Cloud Certified: Professional Data Engineer:** This certification demonstrates your ability to

design, implement, and manage data solutions on Google Cloud Platform.

2. **AWS Certified Data Analytics – Specialty:** This certification validates your ability to design, implement, and manage data analytics solutions on Amazon Web Services (AWS).

3. **Microsoft Certified: Azure Data Engineer Associate:** This certification demonstrates your ability to design, implement, and manage data solutions on Microsoft's Azure platform.

4. **Data Science Council of America (DASCA) Senior Data Scientist (SDS):** This certification is designed to assess your knowledge of data science fundamentals and best practices.

5. **Open Certified Data Scientist (Open CDS):** This certification is designed to assess your ability to apply data science skills to real-world problems.

No matter what your existing skillsets are, or ones you plan to obtain over the next three years as you plot your civilian mission, it's important to hone your skills NOW. I've seen firsthand how expertise in data science, AI, machine learning, and IT can set you apart in today's job market.

Even if you're not planning to dive headfirst into a career specifically in these fields, these skills can be your secret weapon, enhancing your prospects and opening doors to exciting opportunities as you transition into civilian work. It's a wise investment in your career journey.

2.3 Networking 101: Start Growing Your Civilian Network - Internet

As you plan the path from military service to a civilian career, you're stepping into a world where networking is your secret weapon.

It's not just about collecting contacts; it's about cultivating valuable professional relationships that can open doors to exciting job opportunities and offer guidance and support throughout your civilian career.

In this chapter, we'll dive into the digital side of networking, where platforms like LinkedIn, Slack, and Discord can be your allies. But remember, networking isn't confined to the online world, and we'll also discuss strategies for making meaningful connections beyond the phone or computer screen.

LinkedIn: As the world's largest professional networking platform, LinkedIn is an invaluable tool. Start by creating a comprehensive profile highlighting your skills, experience, and career interests. Then, begin to connect with people you know and would like to know. Don't be shy about reaching out to people in industries or roles you're interested in. You'd be surprised at how willing people are to help. Join groups related to your career interests, participate in discussions, and share insightful content. These actions will help you get noticed and build credibility.

Slack and Discord: These platforms are excellent for joining industry-specific or job role-specific communities. Many of these communities are full of professionals who share advice, job listings, and support. It's a great way to learn, network, and show your enthusiasm for your chosen field.

While these platforms are powerful networking tools, it's important not to overlook the potential of networking offline.

Local Meetups and Professional Events: Industry conferences, career fairs, and networking events are all great places to meet professionals in your field of interest.

Consider preparing a short "elevator pitch" referencing your background and career goals for these events.

Executive coach, Dr. Andrea Wojnicki has a simple exercise to keep your elevator pitch current and relevant using the "Present, Past, Future" technique. Remember, stay in the first person, pull on the examples you uncovered in the previous skills and interests exercises and keep it under 15 to 20 seconds.

Present: Who you are, and what you do (highlights).
Past: What you have done in the past (establishing credibility).
Future: What you are looking forward to or working towards (concise, positive, and enthusiastic).

Here are a few examples:

<u>E4-E7 (Enlisted):</u>
Present: *"Right now, I'm serving as a team leader in Houston, where I guide and mentor a group of dedicated individuals to meet our mission objectives."*
Past: *"Throughout my 12-years in the military, I've honed my leadership and problem-solving skills, often in high-pressure situations."*
Future: *"In the civilian world, within the next 2 years, I'm looking to leverage my leadership experience and teamwork abilities in a project management role, ideally in Chicago, where I can drive results and foster collaboration in the manufacturing industry."*

<u>O4-O6 (Officer):</u>
Present: *"Currently, I'm leading a cross-functional team in a strategic planning role based in San Diego, ensuring alignment with our long-term goals."*
Past: *"Over the past 10 years in my military career, I've successfully managed complex projects, including a critical overseas mission that required precise coordination and resource management."*

Future: *"Looking ahead, in the next 2 years, I'm eager to transition my leadership and strategic planning skills to a role in corporate strategy, ideally in New York City, where I can contribute to shaping the future of an organization."*

Don't forget to follow up afterward with the people you meet!

Informational Interviews: Consider reaching out to professionals in your field of interest for an informational interview. This is a casual conversation where you ask for career advice, industry insights, or feedback on your career plans. It's not a job interview, but it's a great way to build relationships and learn. Be clear that your availability to work outside of the military may be several months or years away, and that you seek guidance and practice. NOTE: While your TAPS/ETAPS classes will cover interviews, this is a practice-only experience, and you are seeking constructive feedback on what you did or could do better.

Veteran Networking Groups: Connecting with other veterans who have successfully transitioned can provide valuable advice and insights.
Here's an A-Z list of organizations you may want to check out:
- American Corporate Partners
- American Legion
- Air Force Association
- BootUp PDX
- Code Platoon
- Coast Guard Foundation
- Disabled American Veterans
- Hire Heroes USA
- Hiring our Heroes
- Marine Corps Association Foundation

- Navy League of the United States
- Techstars
- The Mission Continues
- USO
- VFW Foundation
- Veteran Ventures
- Veterans in Transition Career Network
- Veterati
- Warrior to Work

These organizations offer a variety of mentorship programs, including one-on-one mentoring, group mentoring, and online mentoring.

Remember, networking is not just about taking; it's about giving. As you grow your network, look for opportunities to help others. This could be sharing a helpful article, offering to introduce someone to a contact, or giving words of support to someone going through their career transition. The relationships you build through networking are crucial, not just for your transition but for your ongoing career growth.

Start growing your network today. It might feel a bit unfamiliar at first, but with time, it will become second nature. Your network reflects your professional identity. Make it count.

2.4 Networking 101: Start Growing Your Civilian Network – Civic and Volunteer

While technology platforms and professional events are powerful networking tools, never underestimate the potential of civic engagement and volunteer work for building your network.

Participating in local community groups, civic organizations, or volunteer work aligns you with individuals who share similar interests. These connections can lead to unexpected networking opportunities. Additionally, volunteer work allows you to demonstrate your skills, leadership, and team-player qualities outside a traditional work environment.

According to a study by the Corporation for National and Community Service (CNCS), volunteers have a 27% higher likelihood of finding a job after being out of work than non-volunteers. This correlation is even stronger for rural areas and individuals without a high school diploma.

The report suggests that volunteering offers networking opportunities and allows individuals to demonstrate their skills, work ethic, and character. It can also offer a way to expand your technical or functional skillsets.

In the upcoming section of this book dedicated to resume writing, we will delve into the significant role that volunteer experience plays in enhancing your resume and boosting your career prospects. It's worth noting that professionals across industries increasingly recognize the value of volunteer work, with a LinkedIn survey revealing that 41% of respondents consider it as valuable as paid work experience. Furthermore, 20% of hiring managers in the same survey admitted to making hiring decisions based on a candidate's volunteer experience. We'll explore how to effectively showcase your volunteer roles, highlighting the skills and experiences gained in your resume and during job interviews.

Remember, your professional growth is often intertwined with personal growth and community `involvement. Through civic and volunteer work, you can develop new skills, gain unique experiences, meet diverse groups of people, and forge relationships that can provide significant networking opportunities.

So, as you prepare for your transition, consider incorporating civic duties or volunteer work into your networking strategy and your resume!

Types of Networking: Transactional vs. Intentional Networking

Transactional – focus on your goal.
Start every networking opportunity with this question, "What do I get out of this relationship?" Using a purely transactional approach is a great way to start the networking process, however, it's limited in scope. Transactional networking works best when there's money (not salary) at stake, perhaps you're hiring a contractor or buying a car. In today's market, looking for a job can be purely transactional, but in terms of networking for a career well in advance of your availability, people may be turned off by a perceived desire to get what you want quickly with little or no investment in that relationship.

In her online blog, Dynamic Transitions, NY-based psychologist, and executive career coach Lisa Orbe-Austin shared more about transaction-only approaches in Relational v. Transactional Networking: Focus on "Us" Rather than "Me". Orbe-Austin says, "People can sense this a mile away. When you are super busy, like the type of person you want to connect with is, the last thing that you want to do is give away your precious time to someone who feels like they are just interested in taking."

Intentional – Focus on the Person
Jordan Harbinger, creator and podcast host of The Jordan Harbinger Show shared his thoughts on developing intentional relationships while networking. An article in SUCCESS outlines Jordan's beliefs, "As I went about

intentionally building my network, I excluded transactions. I wasn't interested in networking in the same smarmy way that some of us know it. I wanted to build relationships with good people. These relationships could lead to opportunity, prosperity, or flexibility, but that wasn't the point. I led with value. I always try to figure out how I could help others without the expectation of anything in return."

The tagline for his first company, The Art of Charm was, 'Leave everything better than you found it.' He continued, "We wanted something meaningful and actionable that was entirely unambiguous. It meant we weren't playing for the next quarter. We were playing for the next quarter of a century. If I was looking at each interaction from the perspective of how I could give and that I should always leave something better than I found it, it meant I wasn't focused on myself but focused entirely on others. The byproduct of this generosity was that, as time went on, great opportunities came our way."

Perceived Privilege and Networking in Marginalized Communities

Despite significant advancements in the corporate sphere regarding Diversity, Equity, Inclusion, and Belonging (DEIB), perceptions persist that dominant cultural groups, often identified as predominantly white, male, and of higher income, maintain substantial influence over job opportunities through their extensive networks.

Over my long career, I've observed corporations hiring in what I refer to as 'clusters', groups of individuals from similar backgrounds, schools, or regions which can inadvertently shift the balance from inclusion to exclusion. While this may be unintentional or based on shared experiences, it can inadvertently create a homogenous work environment. "Clustering" can lead to reduced diversity, thereby shifting the balance from an inclusive workspace to one that is more exclusive and less representative of the broader community.

Although the frequency of exclusionary practices has declined in the past decade, bias remains an ever-present concern. It's the responsibility of each individual to challenge both their conscious and unconscious biases. Moreover, corporations must cultivate environments that uphold equitable opportunities for all.

Walls of division exist whether you belong to a perceived privileged group or a marginalized one. How can we navigate these barriers? Both marginalized and non-marginalized individuals can benefit from reminders about amplifying voices, fostering allyship, recognizing strengths, and maintaining empowerment.

Empowerment doesn't solely emanate from within. Given these perceptions, it's imperative for marginalized individuals to align themselves with organizations that uphold inclusive policies reflecting their values and aspirations. Additionally, recognizing the potential of inclusive networking groups can significantly aid one's professional journey. By affiliating with these groups and capitalizing on their networks, one can adeptly traverse the professional realm, unlocking opportunities in line with one's goals.

Interestingly, the U.S. military historically boasted a greater male presence than female, though this discrepancy is diminishing. According to the Department of Defense's 2020 data, women constituted about 16% of the active-duty military. While racial and ethnic diversity in the military mirrors that of the broader U.S. populace, statistics denote approximately 70% identify as White, 17% as Black or African American, 12% as Hispanic or Latino, 5% as Asian, and a combined 5% for other races and ethnicities.

This section strives to critically assess the perceived benefits ascribed to 'privileged' groups, understand how these advantages might be construed as exclusionary by some, and offer insights on promoting inclusivity.

If you consider yourself part of a privileged group and wish to be an advocate for inclusivity, contemplate the following:

• Cultivate Awareness: Recognize the nuances of 'privilege' and ascertain where you stand within this framework. When networking, be cognizant of potential power dynamics and engage with marginalized individuals respectfully, inclusively, and without preconceived notions.

Here are actionable steps to be aware of privilege during networking:

• Acknowledge Biases: Be conscious of and confront your inherent biases.

• Respect Boundaries: Respect personal boundaries and never coerce someone into sharing personal stories.

• Active Listening: Give undivided attention when others speak and avoid interrupting.

• Offer Support: Be an advocate and a pillar of encouragement.

• Amplify Voices: Use your position to elevate the voices of marginalized individuals.

Remember, networking is reciprocal. It's about both offering and receiving. Make it a point to share your expertise, resources, and support.

By acknowledging privilege and its associated dynamics, we can contribute to a networking ambiance that champions inclusivity. Take on the role of an ally, backing the ambitions and voices of marginalized individuals.

Empowering Yourself and Enhancing Networking as a Marginalized Individual:

• Self-Awareness: Recognize your strengths and unique value propositions.
• Confidence Building: Nurture self-belief and acknowledge your potential.

- Storytelling: Narrate your experiences compellingly, accentuating your strengths.
- Professional Development: Pursue opportunities for growth in your field.
- Networking Forums: Attend industry-specific networking events, both virtual and physical.
- Online Presence: Leverage platforms like LinkedIn for networking.
- Elevator Pitch: Develop a concise, impactful summary of your professional story.
- Mentorship: Seek guidance from experienced mentors.
- Self-Advocacy: Represent your needs confidently.
- Building Relationships: Cultivate meaningful professional relationships.
- Resource Utilization: Exploit resources tailored for marginalized communities.
- Resilience: Maintain perseverance in the face of challenges.
- Support Networks: Establish a reliable support system.
- Community Involvement: Engage in industry-related community activities.
- Seek Allies: Collaborate with those dedicated to diversity and inclusion.

Rakshitha Arni Ravishankar, a renowned business writer, emphasizes in her numerous articles for Harvard Business Review and Ascend, the importance of recognizing how distinct attributes, like ethnicity, economic status, or cultural background, can be assets in the corporate domain. For instance, an immigrant with entrepreneur parents likely possesses multilingualism, cultural adaptability, and versatile business insights. Such skills, honed over a lifetime, can distinguish you professionally.

Ravishankar suggests introspecting on questions like: What strengths define me? What unique skills have my experiences endowed me with? How can I frame my narrative to manifest as the ideal fit for a specific role or promotion?

If you are from different cultural, ethnic, and socioeconomic background, realize that you bring a wealth of diverse experiences, insights, and problem-solving skills. Your multicultural environment has gained you entry into a world others rarely see, and an innate ability to navigate and bridge cultural differences. As corporations lean to globalization and global networks, this is an increasingly valuable skill.

More importantly, your unique perspectives and life experiences have the opportunity to challenge the status quo and drive innovation within organizations. Diversity of thought leads to more robust discussions, out-of-the-box thinking, and well-rounded decision-making. Ask any 'design thinking' expert the secret to success in groups? Difference of perspective. When people with different viewpoints collaborate, they can uncover blind spots and biases, leading to more inclusive and effective strategies.

In this context, networking also takes on a new dimension. It becomes not just about forming connections for personal advancement but about building a diverse network that can offer a range of viewpoints and experiences. This form of networking enriches anyone's professional journey, providing insights and knowledge that might not be accessible in more homogenous circles.

Recognizing and harnessing the power of your unique background is not just beneficial for you alone; it's imperative for the growth and success of an organization as a whole. The goal? Inclusivity leads to growth and professional development and networking and gaining perspective within this context paves the way for a more dynamic, innovative, and equitable corporate culture.

2.5 Professional Development: Civilian Certifications, Training, and Skillbridge Programs

As we covered earlier in Chapter 2, obtaining certifications in areas like AI, cloud computing, data science, and IT can significantly boost your qualifications for civilian careers.

However, the world of professional development is vast, offering a multitude of opportunities beyond certifications. In addition to these specialized certifications, there are various other pathways to enhance your skill set and transition smoothly into civilian employment.

Keep in mind that several resources provide free or affordable training programs, making it easier for you to acquire new skills. We'll also explore initiatives like SkillBridge, which bridges the gap between your military service and a civilian career, as well as apprenticeship and training programs designed to further your professional growth.

Let's drill down further:

1. **Identify Your Career Goals:** You've started the journey to identify your specific skills and interests and have a few industries and job roles that interest you. Go back to your list and seek out the skills and qualifications they require.

2. **Research Certifications:** Identify certifications that align with your career goals. Consider their credibility, industry recognition, and relevance to your desired job role.

3. **Plan Your Time:** Certifications take time. With your three-year head start, plan when and how you'll complete these courses.

4. **Budget:** Some certifications have associated costs. Consider these in your transition budget.

5. **Get Certified:** Complete the coursework and pass the required exams to earn your certification.

6. **Update Your Resume and LinkedIn Profile:** Once you've earned a certification, be sure to add it to your resume and LinkedIn profile.

Here's a list of valuable civilian certifications, both free (no-cost) and paid (fee-based), across various fields:

Marketing
Free:
Google Digital Garage: Fundamentals of Digital Marketing
HubSpot Academy Inbound Marketing Certification
SEMrush Academy: SEO Fundamentals Certification
Microsoft Learn: Azure Fundamentals
Google Digital Marketing & E-commerce Professional Certificate
Salesforce Administrator Certification

Paid:
HubSpot Academy Sales Hub Certification
HubSpot Academy Marketing Hub Certification
Google Ads Certification
Facebook Ads Blueprint Certification
LinkedIn Ads Certification
Hootsuite Certification
Mailchimp Certification

Cloud
Free:
Google Cloud Platform Fundamentals
Amazon Web Services Cloud Practitioner
Microsoft Azure Fundamentals
Paid:
Google Cloud Certified Professional Cloud Architect
Amazon Web Services Certified Solutions Architect - Professional

Microsoft Azure Solutions Architect Expert

IT
Free:
CompTIA A+
CompTIA Network+
CompTIA Security+
Paid:
Cisco Certified Network Associate (CCNA)
VMware Certified Professional - Data Center Virtualization
(VCP-DCV)
Microsoft Certified: Azure Solutions Architect Expert

AI
Free:
Google AI Fundamentals
Amazon Machine Learning Fundamentals
Microsoft Azure AI Fundamentals
IBM Cognitive Class: Data Science Fundamentals
Paid:
Google Cloud Certified Professional Cloud Machine Learning
Engineer
Amazon Web Services Certified Machine Learning Specialist
Microsoft Certified: Azure AI Engineer Associate

Data
Free:
Google Cloud Platform Big Data Fundamentals
Google Analytics Individual Qualification
Amazon Web Services Certified Data Analytics - Specialty
Microsoft Certified: Azure Data Fundamentals
Paid:
Google Cloud Certified Professional Data Engineer
Amazon Web Services Certified Solutions Architect -
Professional
Microsoft Certified: Azure Data Engineer Associate

Operations
Free:
APICS Certified in Production and Inventory Management (CPIM)
ASQ Certified Quality Engineer (CQE)
Project Management Institute (PMI) Project Management Professional (PMP)**
Paid:
APICS Certified Supply Chain Professional (CSCP)
ASQ Certified Six Sigma Green Belt (CSSGB)
Project Management Institute (PMI) Program Management Professional (PgMP)
Certified Financial Planner (CFP)
**does not include exam fee

Project Management
Free:
Project Management Institute (PMI) Project Management Professional (PMP)**
CompTIA Project+
Google Project Management Professional Certificate
Paid:
Project Management Institute (PMI) Program Management Professional (PgMP)
Project Management Institute (PMI) Agile Certified Practitioner (PMI-ACP)
Scrum Alliance Certified ScrumMaster (CSM)
**does not include exam fee

Cyber
Free:
CompTIA Security+
EC-Council Certified Ethical Hacker (CEH)
ISACA Certified Information Systems Auditor (CISA)
Paid:
Offensive Security Certified Professional (OSCP)
Certified Information Systems Security Professional (CISSP)

Military Education Benefits

It's important to note that as a military service member, you may have access to education benefits that can cover the cost of certifications and training. The Post-9/11 GI Bill, for example, not only covers degree programs but also non-degree programs like certification courses. The Tuition Assistance program also covers vocational and technical training.

1. **Education Service Officer (ESO) or Counselor:** Your ESO or counselor can provide information about the education benefits available to you and how to apply for them. They can guide you through processes like using your GI Bill benefits for certification programs.

2. **Military Branch Education Offices:** Each branch of the military (Army, Navy, Air Force, Space Force, Marines, Coast Guard) has its own education office that can provide information about available education benefits and resources.

3. **Department of Veterans Affairs (VA):** The VA provides comprehensive information about education benefits on its website, including the post-9/11 GI Bill, Montgomery GI Bill, and others.

4. **Official Military Websites:** Websites like Military OneSource offer information on education resources, including tuition assistance and scholarship programs.

5. **Online Search:** Simply conducting an online search for "military education benefits for certifications" can lead to a wealth of resources and information.

6. **Career Service Organizations:** Organizations like Act Now Education, ACP or the USO

offer career services for veterans, including education and training resources.

7.　　　　**Fellow Veterans:** Veterans who have already transitioned may provide first-hand advice and share resources they found helpful.

By exploring these sources, you can learn about various education benefits available and how to utilize them effectively. While navigating these resources, remember to verify the information from reputable sources to avoid any misinformation or misinterpretation. The more informed you are, the better decisions you can make about your career transition.

Keep in mind, benefits vary by service branch and your eligibility depends on factors like length of service. Therefore, it's essential to speak with your Education Service Officer (ESO) or counselor within your military branch to understand your benefits and how you can use them to fund your professional development. Planning this early can help you maximize these resources as you prepare for your transition to civilian life.

Exploring DOD Skillbridge Programs

Another excellent resource for transitioning servicemembers is the Department of Defense Skillbridge program. This program connects service members with industry partners in real-world job experiences during the last 180 days of their military service.

Skillbridge offers a wide variety of training, internship, and apprenticeship opportunities across many sectors, including IT, manufacturing, logistics, and many more. This can be a fantastic way to gain valuable civilian work experience, learn new skills, and build a network before you officially transition.

Below are some popular Department of Defense Skillbridge programs:

1. **Microsoft Software & Systems Academy** (MSSA): This program provides transitioning service members and veterans with critical career skills required for today's growing technology industry. The MSSA program provides training in a variety of Microsoft technologies, including Azure, Office 365, and Dynamics 365. The program also includes career counseling and job placement assistance.

2. **Onward to Opportunity** (O2O): O2O is a free, comprehensive career skills program by the Institute for Veterans and Military Families (IVMF) that provides civilian career training. The O2O program offers a variety of courses in business, technology, and healthcare. The program also includes career counseling and job placement assistance.

3. **Veterans in Piping** (VIP): Offered by the United Association of Journeymen and Apprentices, this program provides 18 weeks of intensive, full-time training in welding and pipefitting. The VIP program is a registered apprenticeship program that leads to a journeyman's license in welding or pipefitting. The program is free for veterans and includes all necessary tools and equipment.

4. **Troops Into Transportation** (The CDL School): This program provides training for commercial driver licenses, preparing service members for careers in transportation. The Troops Into Transportation program offers a variety of CDL training programs, including Class A CDL training, Class B CDL training, and tanker endorsement training. The program also includes job placement assistance.

5. **Workshops for Warriors (WFW):** This is a training program for veterans interested in advanced manufacturing trades, including welding and machining. WFW offers a variety of training programs in advanced manufacturing, including a welding training program and a machining training program. The programs are free for veterans and include all necessary tools and equipment.

6. **Tech Qualled:** A unique technology sales training program specifically designed for veterans. The Tech Qualled program provides veterans with the skills and knowledge they

need to succeed in technology sales. The program includes training in sales prospecting, sales qualification, and sales negotiation.

7. **Per Scholas**: This IT program offers courses in IT Support, Cybersecurity, Software Engineering, and more. Per Scholas offers a variety of IT training programs, including a software engineering training program and a cybersecurity training program. The programs are free for veterans and include all necessary tools and equipment.

8. **NC4ME** (North Carolina for Military Employment): Provides a variety of training, particularly in healthcare and manufacturing. NC4ME offers a variety of training programs, including a certified nursing assistant training program and a manufacturing training program. The programs are free for veterans and include all necessary tools and equipment.

9. **ACT NOW Education** is a non-profit organization led by a U.S. Navy Officer, Jai Salters whose goal is to link 1000 transitioning military to jobs by 2027. His non-profit offers free or low-cost certification and education resources to qualifying individuals and veterans, offered through generous donations.

Here are some industry specific DOD approved Skillbridge programs:

Agriculture: National FFA Organization

Aviation: Aircraft Owners and Pilots Association (AOPA)

Education: Teach For America

Energy: American Petroleum Institute (API)

Financial services: Securities Industry and Financial Markets Association (SIFMA)

Hospitality: American Hotel & Lodging Association (AHLA)

Media and entertainment: National Association of Broadcasters (NAB)

Public safety: National Sheriffs' Association (NSA)

Real estate: National Association of Realtors (NAR)

Sales: Sales Hacker

Transportation: American Trucking Associations (ATA)
New Skillbridge programs are added often. Check the
Department of Defense Skillbridge page for more
information.

DOD Skillbridge Authorized Organizations:
    American Council on Education (ACE)
    American Red Cross
    BootUp Veterans
    CareerForward
    Cerner
    CipherTech Academy
    CodeCraft Academy
    CyberPatriot
    Defy Ventures
    Fisher House Foundation
    Hiring Our Heroes
    IBM SkillsBuild
    Institute for Veterans and Military Families (IVMF)
    Mercer Group
    Mission: Readiness
    National Center for Healthcare Leadership (NCHL)
    National Council of State Boards of Nursing (NCSBN)
    National Association of State Workforce
    Agencies (NASWA)
    New America
    Oracle Academy
    OutSystems
    Pathways to Success
    Project Management Institute (PMI)
    Salesforce
    ServiceSource
    Skills for America's Future
    Stand Together
    The Home Depot
    Udacity
    University of Phoenix

UpSkill America
Veteran Employment Through Technology Education
Consortium (VET TEC)
Veterans in Piping (VIP)
Workshops for Warriors (WFW)

Learn more on the DOD Skillbridge website or locate your local Skillbridge coordinator through your unit commander or military support services on base.

Remember, each program offers unique opportunities. Be sure to explore each to see which aligns best with your career aspirations and goals. By taking advantage of these resources, you can make significant strides toward your civilian career while still in service. These programs can also be a great way to try out potential career paths and industries before you commit.

As I close this chapter, I'd like to pause and reflect on your journey so far. In Chapter 1, "The Journey Ahead - What to Expect," we peeled back the layers of the civilian job market. We dissected typical hurdles many people, like you, will face when leaving your military career for the corporate or civilian job landscape.

In Chapter 2, "Laying the Groundwork - 3 Years Out," we pointed your compass toward prep mode, taking a deep dive into understanding your strengths outside the uniform and zeroing in on those skills and passions that will make you shine in the civilian world.

We covered Tech, AI, and Data Science, all sectors that are ripe with opportunities for those who are interested in learning more. Networking was covered and we talked about building bridges, both professionally and within communities. And when it comes to training? You, with your military background, are no stranger to that. So, adding some tailored certifications and training to your skill set? That's just refining what you already do best!

As we segue into Chapter 3, we're stepping into the foundational phase of your second year of planning. My goal is to ensure you've got the know-how to make this transition as seamless as possible. The post-military career horizon is bright, and I'm genuinely excited to see where this journey takes you.

Ready to forge ahead?

Let's do this!

CHAPTER 3:
STRATEGIZING YOUR APPROACH - 2
YEARS OUT

3.1 Setting Clear Career Goals

After devoting the first year of your three-year transition journey to laying the groundwork—identifying transferable skills, exploring potential career paths, building your network, researching, and obtaining valuable certifications—it's now time to strategize your approach for the upcoming years. The core of this strategy involves setting clear, precise, and achievable career goals.

Career goals are your guiding light. They give you a direction to follow, benchmarks to measure your progress, and a clear vision of what success looks like for you. Setting clear career goals is not merely stating, "I want to work in tech" or "I want to have a job in project management." Instead, it's about defining what specific job roles you aim for, building on the skills and interests exercise you wrote in the previous chapter by adding new skills you need to acquire. It's also about the companies you're researching and later will be targeting, and what steps you'll take to get there.

When setting your career goals, let's use the SMART framework:

SMART is an acronym used to guide the setting of objectives, particularly in project management and personal development. Here's what each letter in the acronym stands for:
S - Specific:
The goal should be clear and specific, allowing you to understand exactly what is expected.
Instead of "I want to be healthier," a specific goal might be "I want to run a 5K in under 30 minutes."

M - Measurable:
You should be able to measure progress towards the accomplishment of the goal.
By determining metrics or indicators, you can track your progress and know when the goal has been achieved.
Using the above example, the 5K race and the 30-minute timeframe provide measurable metrics.

A - Achievable (or Attainable):
The goal should be realistic and attainable given the available resources and time.
It shouldn't be too easy that it doesn't challenge you, nor too hard that it's impossible to achieve.

R - Relevant (or Realistic):
The goal should matter to you or align with other relevant goals.
It's about ensuring that the goal fits with your broader career or personal aspirations.
For example, if your broader goal is to improve your physical health, training for a 5K race is relevant.

T - Time-bound (or Time-limited):
The goal should have a deadline or a specific timeframe for completion.
This creates a sense of urgency and helps you prioritize and manage your time to achieve it.

In our example, setting a deadline like, "I want to run a 5K in under 30 minutes by December" gives you a clear time frame.

Consider the following statistic: according to a study by the Harvard MBA program, the 3% of graduates who had written career goals ended up earning ten times as much as the other 97% combined, ten years post-graduation.

While earnings aren't the only measure of career success, this illustrates the power of clear, written career goals.

Exercise: Crafting Your Career Goals
Objective: To help you define and refine your civilian career aspirations by setting specific, measurable, achievable, relevant, and time-bound (SMART) goals.
Duration: 60-90 minutes
Materials Needed: Notebook or digital device, your interests and skills exercise from the previous chapter, and a calm, quiet space to think and reflect.

Step 1: Reflect on Your Desired Career Path
Revisit the interests and skills exercises you wrote in the previous chapter. Identify two to three fields or roles that most resonate with your passions and strengths. Write them down.

Step 2: Research and Be Specific
For each field or role, conduct a quick online search. What are the sub-specializations or specific job titles in that sector? (e.g., not just "cybersecurity," but "network security analyst" or "ethical hacker"). List down these roles.

Step 3: Add Time Constraints
For each specific role or job title, estimate a realistic timeframe to achieve it, considering the time left in your transition and any additional training or education you might need.

Step 4: List New Skills or Certifications Needed
Based on your research in Step 2, what additional skills or certifications might make you more competitive for that role? Prioritize them based on their importance and relevance to the job. Refer to the Free and Paid Certifications sections of Chapter 2.

Step 5: Break it Down
Break each main goal into smaller tasks or milestones. What steps will you need to take over the next two years to achieve these goals? Be as detailed as possible. This could include things like "Attend a network security workshop in May" or "Connect with three professionals in the AI sector this month."

Step 6: Flexibility Check

Imagine the job landscape changing or your personal interests evolving. Would your goals still stand? Adjust your goals, ensuring there's room for potential shifts, but without losing your core focus.

Step 7: Review and Commit

Review your goals, ensuring they follow the SMART criteria. Once satisfied, make a commitment to yourself. Place your goals somewhere visible – on your desktop, printed above your workspace, or as a monthly reminder in your calendar.

Reflection: Now that you've set these goals, how do you feel? Energized, overwhelmed, or something in between? Remember, the path might be challenging, but with clarity and focus, you're setting yourself up for success. Your military training has prepared you for missions far more daunting than this. Treat this transition as your civilian mission, with a clear objective and a strategic plan to achieve it. And always, remember you're not alone on this journey. Many have walked this path before and succeeded, and so will you.

As you close this exercise, look forward to the upcoming chapters, where we'll provide you with actionable strategies and tools to make your goals a reality. By setting clear career goals now, you're providing a roadmap for the next two years of your transition journey. This roadmap will guide your decisions and help you focus your efforts, enhancing your chances of landing a civilian career that aligns with your aspirations, skills, and values. In the following chapters, we'll delve into specific strategies and steps to help you reach these goals.

Keep moving forward!

3.2 Building a Strong Resume and LinkedIn Profile

A strong resume and a compelling LinkedIn profile are key components of your career transition toolkit. Together, they serve as your professional introduction and make a powerful first impression on potential employers. As career transition experts often stress, the attention to detail in these tools is paramount.

Here are concrete steps to help you build a standout resume and LinkedIn profile:

Step 1: Understand Your Audience

Start by understanding what potential employers are looking for. Review job postings in your target industry, noting the skills and qualifications that are frequently mentioned. My friend and colleague, Marisol Maloney, a veteran, Mil2Civ career coach and 'secret squirrel' recruiter, emphasizes the importance of aligning your resume and LinkedIn profile. "While you will have dozens of resumes based on the jobs you will go after, the foundation of that resume needs to be in your LinkedIn profile. You can always build on that foundation for specific jobs but have the bare bones on your LinkedIn profile so if a recruiter can't find a resume – they have a profile to back it up." Get in the practice of outlining your current military work and update it as you grow in your military career. Keeping your LinkedIn profile up to date is an excellent practice and is an insurance against you leaving out certifications or unique job roles.

Step 2: Craft Your Resume

Your resume should be clear, concise, and tailored to each job you apply for. Start with a strong summary statement that highlights your key skills and career goals. Then, list your work experience, starting with your most recent position. Highlight your accomplishments in each role, quantifying them whenever possible. According to a study by Ladders, recruiters spend an average of just 7.4 seconds reviewing a resume. Make those seconds count with a resume that is easy to read and quickly communicates your value.

Step 3: Optimize Your LinkedIn Profile

Your LinkedIn profile is a digital extension of your resume but with more room for detail and personality. It's also a place where you can get other professionals to advocate for you by writing a review or recommendation. I highly recommend you include a professional photo, crafting a compelling headline, and writing a summary that tells your professional story. Be sure to list all your work and volunteer experiences, skills, certifications, and education. Also, request recommendations from colleagues and superiors to boost your credibility.

NOTE: If you are active duty, it's fine to have a photo in uniform against an American flag. However, if you are in your final 180 days of service, consider investing in a professional headshot, using a photographer or an AI assisted headshot creator that is using 25 or more photos of your face. Keep in mind that many AI generators are producing results that do not represent what you look like. Be discerning and exercise good judgment.

Step 4: Highlight Your Military Experience

Translate your military experiences into civilian terms, focusing on the transferable skills you gained. Leadership, strategic thinking, teamwork, and problem-solving are all highly valued in the civilian world. According to a survey by LinkedIn, 80% of hiring managers believe veterans bring high levels of discipline, problem-solving, and teamwork to their companies.

Step 5: Add Volunteer Experience to Resume and LinkedIn Profile

A LinkedIn survey found that 41% of professionals surveyed considered volunteer work as valuable as paid work experience. Furthermore, 20% of hiring managers in the survey reported that they had made a hiring decision based on a candidate's volunteer work experience.

Interestingly, this same survey, conducted in 2011 included nearly 2,000 professionals in the United States. The survey found that 89% of respondents had personally had experience volunteering, but only 45% of respondents included their volunteer experience on their resume!

If you have volunteer experience, be sure to include it on your resume and highlight the skills and experience that you gained from your volunteer work in your cover letter and interviews.

Here are some tips for highlighting your volunteer experience on your resume and in your job search:

• List your volunteer experience in the same section as your paid and or military work experience.

• Include the name of the organization, your role, and the dates you volunteered.

• Be specific about the tasks and responsibilities you had in your volunteer role.

• Highlight any skills or experience that you gained from your volunteer work that are relevant to the jobs you are applying for.

• Be prepared to talk about your volunteer experience in job interviews. Be able to articulate the skills and experience that you gained from your volunteer work, and how those skills and experience make you a good fit for the job you are applying for.

• Quantify your results using numbers (#), percentages (%) and dollars ($) where possible.

Here are some examples:
Enlisted:
• Volunteered at a food bank and helped to distribute food to 100 families each week, totaling over 5,000 meals served.
• Mentored an at-risk youth through a Big Brothers Big Sisters program for 2 years, helping him to improve his grades and graduate from high school.
• Coached a youth soccer team for 1 season, leading them to a winning record and a championship.

- Led a team of 10 volunteers to build a new playground for a local park, saving the community over $10,000 in construction costs.
- Organized a fundraiser for a local charity and raised over $1,000, which was used to purchase new medical equipment for a children's hospital.

Officer:
- Managed a team of 50 volunteers to provide disaster relief to victims of a hurricane, helping to rebuild over 100 homes and distribute food and water to over 1,000 people.
- Developed and implemented a new volunteer training program for a local nonprofit organization, increasing the number of trained volunteers by 25%.
- Served as the board president of a local homeless shelter for 3 years, overseeing a budget of $1 million and providing shelter and services to over 500 homeless people each year.
- Represented a nonprofit organization at a state-wide conference, speaking to over 1,000 people about the importance of volunteering.
- Secured a $10,000 grant for a local charity, which was used to fund a new job training program for low-income individuals.

Step 6: Update Regularly

Ensure your resume and LinkedIn profile are free of typos and grammatical errors. Also, regularly update them with any new skills, experiences, or achievements.

Building a strong resume and LinkedIn profile is a process that requires time and attention to detail. But the investment is well worth it. These tools are often your first chance to show potential employers what you bring to the table. Make sure they present you in the best possible light. As you move further along your transition journey, your resume and LinkedIn profile will continue to evolve, just like your career goals. Stay committed to refining and updating them, and they'll serve you well in your civilian career.

Federal Resume Writing and USAJOBS.gov

When applying for federal jobs, it's important to note that the resume requirements are different from those in the private sector. A federal resume is typically longer and includes more details about your skills, tasks, and accomplishments related to the job you're applying for.

USAJOBS, the federal government's official job site, offers a comprehensive Federal Resume Writing Guide. This guide includes step-by-step instructions and tips to help you create a strong federal resume.

Some key points from the guide include:

1.   **Job Announcement:** Tailor your resume to the job announcement's requirements. Include specific examples of how you meet the qualifications and requirements listed in the announcement.

2.   **Length:** While private sector resumes are often recommended to be no more than two pages, federal resumes can be much longer depending on your experience and the position for which you're applying.

3.   **Details:** Provide more specific details about your duties and accomplishments in each role. This includes percentages, dollar amounts, and other quantifiable achievements.

4.   **Keywords:** Incorporate keywords from the job announcement to help your resume get past the initial screening process.

Remember, writing a federal resume requires a different approach. Take advantage of the resources provided by USAJOBS.gov to create a resume that aligns with federal hiring practices.

Let's delve into the major differences between civilian and federal resumes:

1. Length: A civilian resume is typically 1-2 pages long, whereas a federal resume can be much longer. Federal resumes often range from 3-5 pages or more, depending on the applicant's experience and the job's requirements.

2. Detail: A civilian resume usually provides a snapshot of your experience and skills. It focuses on highlights and major achievements. On the other hand, a federal resume requires more detailed information about each job you've held. It typically includes hours worked per week, supervisor contact information, and a comprehensive description of duties, achievements, and related skills.

3. Format: Civilian resumes often have more flexibility in terms of format and design. They can include a wide variety of headings, layouts, and even visual elements, depending on the industry. However, federal resumes follow a more standard format, typically a chronological layout detailing your work history beginning with your most recent job.

4. Tailoring: While both civilian and federal resumes should be tailored to the job description, a federal resume should closely mirror the specific language and requirements found in the job announcement.

5. Objective/Summary: A civilian resume might start with a career objective or summary statement. A federal resume, instead, starts with the job announcement number, job title, and series and grade, which are specific details relating to the federal job vacancy.

6. KSAs: Federal resumes often require the inclusion of KSAs (Knowledge, Skills, Abilities). This section requires detailed narrative statements related to the specific knowledge, skills, and abilities required for the job.

7. Personal Information: Federal resumes often require more personal information than civilian resumes, including citizenship, security clearance, and veteran's preference. Note: Security Clearances can be listed on civilian resumes for DOD contractor type positions.

These are some of the key differences between civilian and federal resumes. As you're now at the two-year mark in your transition journey, it's essential to understand these differences and start crafting both types of resumes accordingly. While it's beneficial to start early on constructing your resumes, remember that you should align the timing of your job applications with your transition timeline.

Here's a summary of what you've learned and accomplished so far in the process:

1. **Understanding the Civilian Job Landscape:** You've spent time gaining insights into the civilian job market, learning about the key industries, emerging sectors, and most in-demand skills.

2. **Identifying Transferable Skills and Interests:** You've begun the process of translating your military skills into civilian terms, identifying your core competencies, and matching them with potential careers.

3. **Exploring Initial Career Path Considerations:** You've started researching potential careers and industries of interest, including an emphasis on new technical careers for non-technical individuals.

4. **Building Your Network:** You've begun the task of expanding your civilian network, both online and offline. This includes leveraging platforms like LinkedIn, participating in networking events, and engaging in community service and civic work.

5. **Professional Development:** You've explored a list of free and paid civilian certifications and started pursuing those relevant to your chosen career path. You've also discovered resources like the DOD Skillbridge programs, and the Hiring our Heroes initiative through the US Chamber of Commerce.

6. **Setting Clear Career Goals:** You've set specific, realistic, and time-bound career goals, creating a roadmap for the next phases of your transition journey.

7.   Building a Strong Resume and LinkedIn Profile:
You've started to build your civilian and federal resumes and LinkedIn profile, learning to highlight your skills and achievements effectively.

Remember, it's crucial to appreciate the timeline and avoid jumping ahead. While the urge to apply for positions may be strong, acting too soon may lead to opportunities that are out of sync with your anticipated transition date.

Instead, use this time wisely to further solidify the foundations you've been building over the past year. Continue growing your professional network, both in person and through online platforms such as LinkedIn. Keep engaging with professionals in your desired field, attend industry events, and stay active in relevant online communities.

During your second year, continue expanding your understanding of both civilian and federal job application processes. Refine your resumes—both federal and civilian—ensuring they're tailored to the types of roles and organizations that align with your career goals.

Remember the importance of lifelong learning. If there are additional certifications, training, or courses that could increase your competitiveness, consider pursuing them now.

Maintain your focus on defining and redefining your career goals, as these will continue to guide your decisions and actions in the coming phases of your transition.

As you progress, always carefully read the job descriptions. When you are closer to your application time, remember to follow the application instructions to ensure your resume meets all requirements. If you need help, you can seek the guidance of the staff at any family and career readiness office on base or check into the many free volunteer resume writing services in the veteran non-profit spaces.

Even though you are more than 1 year out from your availability, let's take a few minutes to do a deep dive into refining a resume using ChatGPT.

Exercise: Refining Your Resume with AI Assistance
Objective: To align your resume with a specific job description, ensuring relevancy and clarity.
Materials Needed:
Your current resume.
A job description of a position you're interested in (even if it's hypothetical for now).
Access to ChatGPT or another AI platform of your choice.
Steps:
1. **Review and Reflect:** Before diving into the AI platform, read through your current resume. Note down the main points about your military experience that you feel might be transferable to the civilian job you're eyeing.

2. **Job Description Analysis:** Now, read the job description thoroughly. Highlight keywords and required qualifications. Think about how your military experience can be presented in a way that addresses these points.

3. **Engage with the AI:** Go to the ChatGPT platform and input both your resume and the job description. Use the prompt, "rewrite of resume against job description." Let the AI generate a revised resume for you.

4. **Review the AI-Generated Resume:** Go through the revised resume provided by the AI. Does it capture the essence of what you offer? Have essential points been omitted or misrepresented?

5. **Fact Check:** Ensure all information on the revised resume is accurate. Remember, while AI can assist in restructuring and aligning your resume, it might not always capture the nuances or specific accomplishments you've had in your career. And, because it is a machine learning mechanism, it may place details in the wrong place. Do not copy and paste without a full fact and reality check!

6. **Seek Feedback:** Once you're satisfied with the AI-revised resume, consider sharing it with a trusted colleague or

mentor. Getting an external perspective can provide valuable insights. Remember the resources mentioned earlier, like the staff at any family and career readiness office or volunteer resume writing services in the veteran non-profit spaces.

7. Reflect and Iterate: Based on the feedback you receive, and your own insights, iterate on your resume. Remember, a resume is a living document; as you grow and gain more experience or skills, you'll want to update it.

Endnote: Leveraging technology, especially AI tools like ChatGPT, can significantly streamline your transition journey. While these tools offer advanced capabilities to help reorganize and refine your resume against specific job descriptions, always remember that the real essence of your journey, skills, and experiences can't be completely encapsulated by algorithms. Your personal touch, combined with the technological advantages AI offers, will shape the most compelling and impactful resume.

Using ChatGPT to prompt a "rewrite of resume against job description" is an excellent way to ensure alignment and relevancy between your skills and the job's requirements. However, always be mindful that AI is a computer-generated model. While it can guide and suggest changes, the final responsibility of accuracy and authenticity rests with you. Thorough fact-checking is essential.

As you navigate this process, patience and strategic planning are key. This is a period of preparation – and it needs both your enthusiasm and dedication. Remember, every effort you put in now is building momentum, setting you on a strong trajectory for a successful transition into the civilian world.

3.3 Strategic Networking: Making Connections that Count

As you navigate your way through the transition process, one of the key strategies to keep in mind in year two is strategic networking. But remember, at this two-year mark, your networking efforts are not about landing a job immediately. Instead, they are about continuing to build connections, fostering relationships, and setting the stage for opportunities that will come when you are ready to make that leap into the civilian world post-military.

First, let's clarify what strategic networking means in the second year. You've already started this process; you've set up your LinkedIn profile, you've listed your current MOS/Job titles and perhaps also added significant volunteer and civic experience. In short, you've started framing information that could be used as a general resume.

You've worked on your skills and interests to better understand what you want to do when the time comes.

Networking at this stage is still important, but it's not thrusting into high gear quite yet. Do not think you have to start collecting a vast number of contacts or spreading your resume or your interest in future jobs far and wide. Strategic networking involves identifying and connecting with people who are relevant to your specific career interests and can provide valuable insights, guidance, and potentially, opportunities down the line.

Building on what you've learned in the Networking chapter, start by identifying professionals in your desired field. They could be recruiters, HR professionals, industry leaders, and others who could provide valuable insights into your chosen career. Use LinkedIn, as we discussed, to research these individuals and initiate connections. Continue to attend industry events, join professional associations, and participate in online discussions relevant to your career interests.

Importantly, don't underestimate the power of networking through civic and volunteer work. Community service allows you to make meaningful contributions while connecting with individuals from diverse backgrounds. It also allows you to add critical skills to your resume and LinkedIn profile that may not fall in your current or past MOS. Many professionals across the corporate landscape value civic engagement and are actively involved in volunteer work, providing another avenue for connection and conversation.

As you connect with others, think about how you can offer value to them as well. Perhaps you have unique insights from your military experience, or maybe you've recently completed a certification that others might find interesting. By offering something of value, you make the connection mutually beneficial, which strengthens the relationship.

Next, consider informational interviews - conversations where you ask for advice, not a job. Reach out to people who hold positions you're interested in and ask if they'd be willing to share their experiences, insights, and advice about their career path. People generally like to help and share their expertise, especially if it's clear you value their time and insights.

Also, don't forget to tap into your existing network. Former military colleagues who have successfully transitioned to civilian careers can provide invaluable advice and potentially connect you with others in their network.

Finally, keep in mind that networking is a long-term endeavor. It's about building and maintaining relationships over time. So, don't be discouraged if you don't see immediate results. Your networking efforts now are like planting seeds that will bear fruit when you're ready to transition into your civilian career.

So, let's refresh our knowledge so far. What is networking? Networking is about setting the stage for your future job search. It's about making connections that count, which will pave the way for a smoother transition to a civilian career when the time is right. Keep investing in these relationships, nurturing them with regular check-ins and updates about your journey. When you are ready to make that leap, you'll have a robust network supporting you.

According to a LinkedIn study, 85% of jobs are filled via networking. This emphasizes the importance of networking when searching for job opportunities.
A survey by the Adler Group revealed that 46% of men and 39% of women found their current positions through networking, significantly higher than those who found jobs through job boards or other methods.

The U.S. Bureau of Labor Statistics reports that people who use multiple job search strategies have more success in finding jobs than people who use only one.

Strategies to Gain Advantage:
- **Create or Join a Thought-Leadership Group:** Connect with others who are at the same stage in their transition journey. You can do this by starting a thought-leadership group where each member can share ideas, advice, and experiences. This kind of group can be formed in various ways. For instance, it could be a virtual meet-up on a platform like Zoom, a private group on LinkedIn, or a dedicated channel on Slack or Discord. Regular meetings or discussions can foster a strong sense of community and shared purpose, providing support and motivation as you navigate the transition together. Finding these peers might be a challenge at the two-year-out phase, but you can leverage various resources to make these connections.

1. You could reach out to your existing military network to find out if anyone else is planning a similar transition timeline.

2. Consider participating in online forums or social media groups for transitioning military personnel where you can meet others at the same stage.

3.   Transition Assistance Programs (TAPs) and ETAPS, while typically done in the final year, may also offer online resources or communities where you can connect with other transitioning personnel earlier.

- **Build a Personal Brand:** Even if you're not actively job-seeking, start building your brand. A strong personal brand can help you stand out and make a lasting impression. This could involve speaking at industry events, writing articles or blog posts on topics you're knowledgeable about, or sharing insightful content on LinkedIn. Linda Citroen, a personal branding expert and author of the book "Your Next Mission: A Personal Branding Guide for the Military-to-Civilian Transition," emphasizes the power of personal branding for military personnel transitioning into civilian careers. Citroen's book provides valuable insights and practical advice on how to leverage your brand to make a successful transition. By investing time now to build and polish your brand, you are positioning yourself ahead of those who will begin this work only a year in advance of their transition.

- **Get Certified:** Identify certifications or segments of training relevant to your target industry and complete them or create a timeline for completion that you can list on your resume as "in process." Make sure you are signed up for the cert before listing it on your resume. Completing these certifications before your availability will give you a competitive edge over others who are just beginning their preparation a year in advance.

- **Volunteer or Intern:** Seek out volunteer or internship opportunities in your target field. This can give you relevant experience, skills, and networking opportunities. It's a unique approach that few think of employing and can greatly enrich your understanding of the civilian job landscape.

- **Seek Mentorship:** Find a mentor who has already navigated the military to civilian transition. They can provide guidance and support throughout your journey. This relationship, built over two years, will be far more beneficial than rushing to find advice during the last stages of your transition. We've covered mentorship in the earlier chapters. Here are several more resources available to help you find a mentor for your military-to-civilian transition.

1. **American Corporate Partners (ACP):** ACP offers a nationwide mentoring program that connects post-9/11 veterans with corporate professionals for year-long mentorships.

2. **Veterati:** This digital platform facilitates mentorship conversations between veterans, military spouses, and successful industry professionals.

3. **Hire Heroes USA:** This organization provides transition assistance, including one-on-one mentoring and coaching, to military members, veterans, and their spouses.

4. **Hiring our Heroes:** Part of the US Chamber of Commerce, this popular program connects 180 day service members a straight connection to hundreds of companies, streamlining the process for both Mil2Civ and businesses. Their MilSpouse career services and Veteran Mentor network is extensive.

5. **LinkedIn's Veteran Mentor Network:** This is an active group of veterans and supporters who engage in thoughtful discussion and provide advice and guidance for transitioning military personnel.

Remember, finding the right mentor can take time, so it's important to start early. Reach out to potential mentors, and don't be afraid to ask for advice or guidance. Most people are willing to help, especially when they see you're proactive and committed to your career transition.

Ensure you communicate openly with your mentors about your transition timeline and your job availability. Keeping them updated about where you are in your transition journey helps them provide the most relevant guidance and support. And when the time comes, they may be able to connect you with job opportunities or recommend you to their professional network.

By maintaining open communication and building strong relationships with your mentors, you create a supportive network that can greatly aid your successful transition into a civilian career.

Before we move on to the next chapter, let's take a moment to review what you've learned in Chapter 3, "Strategizing Your Approach - 2 Years Out."

This chapter marks a crucial juncture—two years before your transition to civilian life. You've done the groundwork, and now it's time to get strategic with your plan. It begins with setting clear career goals, meticulously designed using the SMART framework. You've become specific, ensuring your objectives are Specific, Measurable, Achievable, Relevant, and Time-bound. Your goals have evolved from vague aspirations to targeted roles, companies, and the steps you'll take to get there.

You pondered the powerful statistic from the Harvard MBA program, which underlines the significant impact of written and well-defined career goals. It's not just the financial gains that matter but setting a path to satisfaction and purpose in your post-service career.

In our practical exercises, you sat down to craft these goals. You revisited your skills and interests, conducted research, added timeframes, identified new skills to learn, and broke it all into manageable tasks. You've accepted the need for flexibility, acknowledging that as the job landscape changes, so may your targets.

When it comes to presenting yourself, you've tackled the challenge of building a foundational resume and a persuasive LinkedIn profile. You've taken advice from veterans like

Marisol Maloney and learned to translate your military experiences into civilian language, ensuring your profile is always current.

You then navigated the distinct world of federal job applications, recognizing the need for a different resume strategy that aligns with federal hiring practices.

Lastly, you've delved into the world of strategic networking. You're not just collecting contacts; you're cultivating relationships, engaging with industry influencers, and carving out your personal brand. Networking has become not just an action but a habit for you—a consistent cultivation of connections that will sustain your career long after you've transitioned.

As you turn the page from this chapter, remember that every step you've taken is laying down the path toward a smooth change into civilian employment. Your military discipline has prepared you for challenges much steeper than this. Carry that determined, strategic approach into the next chapter, where you'll begin to put these detailed plans into action.

Keep moving forward!

CHAPTER 4:
GAINING MOMENTUM - 1 YEAR OUT

4.1 Preparing for the Job Search: Cover Letters, Interviews, and More

Chapters 1 through 3 have been all about laying the foundation for a successful military-to-civilian transition. We started by understanding the civilian job landscape and identifying common challenges faced by transitioning personnel. You have learned about the importance of starting this journey early, preferably three years out, to give you ample time to strategize and prepare.

In Chapter 2, we dove into the groundwork, identifying your transferable skills and interests. We've explored initial career paths, including new technical careers for non-technical individuals. We discussed the importance of growing your civilian network, using both online platforms and offline opportunities, such as civic and volunteer work. We delved into professional development, exploring free and paid civilian certifications and training.

Chapter 3 was all about strategizing your approach. You learned how to set clear career goals and began to build a strong resume and LinkedIn profile. We emphasized the importance of strategic networking, not to land a job immediately, but to build connections that will bear fruit when the time is right.

Now, we're at Chapter 4, Gaining Momentum - 1 Year Out, and it's Go-Time. You're one year away from the transition, and it's time to prepare for the job search.

By making it this far, you've not only laid an unshakable foundation, but you've also positioned yourself two years ahead of the curve. While others may be just beginning to consider their post-military career, you have a two-year head start in understanding the landscape, developing skills, and networking. This foresight and preparation have set you apart, giving you a strategic advantage in the competitive civilian job market. You are now part of an elite group who didn't just wait for transition to become imminent but actively planned for it. As you step into the next phase, remember that this diligence is what differentiates a successful transition from an uncertain one. Let's keep this momentum and take it to the next level.

In Chapter 4, we'll focus on:

Crafting Cover Letters: We'll delve into writing persuasive cover letters that highlight your skills, experiences, and fit for the job.

Preparing for Interviews: We'll discuss strategies to excel in various types of interviews, from traditional one-on-one settings to phone and video interviews.

Salary Negotiation: Learn about salary expectations in the civilian job market and how to effectively negotiate a fair compensation package.

Continued Networking: Keep expanding your network strategically, aiming now to turn these connections into potential job opportunities.

Job Application: Start gearing up to apply for jobs. Keep in mind that the timing of your applications is crucial. Although you're a year away from being available for employment, you should plan your job applications according to your availability. The process of applying, interviewing, and receiving a job offer can take several weeks or even months, depending on the industry and the company.

As a corporate recruiter, I cannot emphasize enough the importance of having a strategic application timeline. Each industry operates at its own rhythm when it comes to hiring and being unaware of when to hit the 'apply' button can lead to missed opportunities or unnecessary stress. It's vital to align your application efforts with industry standards and company hiring cycles to ensure your candidacy is considered at the right time.

As a rule of thumb, consider starting your active job search, or application process and hitting the 'apply' button no earlier than 180 days, but no later than 90 days from your availability. This allows adequate time for potential employers to process your application and schedule interviews while ensuring that you're ready to start work when the job offers start coming in.

Exercise: Mapping Out Your Application Timeline
Objective:

This exercise is designed to help you visualize the timeline and potential challenges of job applications, underscoring the importance of timing your applications effectively.
Instructions:

Reflect and Write: Start by listing the industries or sectors you selected in Chapter 2. For each industry, write down the average time you believe it takes from application to job offer based on your research or information from your network.

Timeline Creation: On a sheet of paper or digitally, create a 12-month timeline, marking your end point (the date you're available for civilian employment). Label this as "Day of Availability."

Working Backwards: From your "Day of Availability," count and mark 90 days before and then 180 days before. Label these two points as "Optimal Application Start" and "Earliest Application Start" respectively.

Industry-Specific Adjustments: Referring to the list from step 1, adjust the application start dates (both optimal and earliest) for each industry. If an industry generally takes longer to process applications, consider moving the application start date even earlier. Make notes on any industries that may require a different approach.

Reflection: Imagine you applied for a job today, a full year before your availability. Write a short paragraph detailing potential challenges or scenarios you might face by applying too early. Consider factors like employers questioning your availability, juggling multiple job offers that have different start dates, or potential missed opportunities as you wait.

First-Person Affirmation: Conclude this exercise by writing the following affirmation: "I understand the importance of timing in my job application process. By aligning my applications with my availability, I increase my chances of a smoother transition and reduce unnecessary frustrations."

Remember, this exercise is to emphasize patience, strategy, and ensuring that you're not putting the cart before the horse. Getting ahead of yourself might feel proactive, but it's essential to move at a pace that aligns with your overall transition strategy. This careful planning ensures you're not only finding the right opportunities but also that you're poised to seize them at the perfect time. The upcoming sections will provide you with further insights and tools to navigate this important phase of your career transition journey.

Transitioning from a military to a civilian career comes with its unique rhythm, often dictated by the military's structured timelines. Unlike traditional job changes where you might decide when to shift roles or industries, your transition out of active duty is typically determined by the military itself. This can mean you don't have full control over when you become available for civilian employment.

The application, interview, and job offer processes can differ vastly across industries and specific companies. In government contracting, roles often hinge on distinct contracts, filled based on immediate contract necessities. So, while the demand might always be present, the positions may not be long-lasting. Jumping into this too early might leave you waiting.

Conversely, sectors like manufacturing, retail, or the corporate world usually have a consistent demand for certain roles, which allows for a more predictable hiring window. Tech companies, especially startups, can have dynamic hiring needs, adapting quickly to market trends and their growth stages. The finance sector, with its banks or investment firms, often has structured annual recruitment cycles.

Other industries like healthcare and education also follow their specific timelines based on institutional and academic needs. However, these broad strokes don't fit every scenario, and deviations are expected.

To navigate these complexities, it's crucial to dig deep into your target industries. Comprehensive research, combined with leveraging your network—particularly those already in your desired field—can give invaluable insights into hiring timelines.

Remember, the ultimate goal isn't just about securing a position. It's about finding a job that matches your skills, interests, and long-term ambitions, all while fitting into the pre-determined transition timeline you're working within. With patience, networking, and a laser focus on your transition strategy, you're paving the way for a prosperous and meaningful civilian career.

4.2 Targeting Specific Companies and Industries

Researching and targeting specific companies and industries is a crucial part of your job search strategy. A study by the Corporate Executive Board found that candidates who take the time to research companies and tailor their applications are 2.4 times more likely to be shortlisted for an interview.

There are a few steps to effectively research and target companies:

1. Identify Your Target Industries: Based on your skills, interests, and career goals, identify the industries that align with your objectives.

2. Shortlist Companies: Within these industries, shortlist companies that interest you. Look at factors such as company culture, values, products or services, reputation, and opportunities for growth and development.

3. Deep-Dive Research: Conduct deep-dive research on each shortlisted company. Go beyond their website and LinkedIn page. Look at news articles, press releases, annual reports, and employee reviews on sites like Glassdoor.

4. Understand their Needs: Try to understand what the company might need in an employee. Look at their job postings to get a sense of the skills and qualifications they value.

5. Tailor Your Applications: When the time comes to apply, tailor your resume and cover letter to highlight how your skills and experiences align with the company's needs.

To help you stay organized and track your research, consider creating a research journal or a spreadsheet. For each company, jot down key facts, your impressions, questions you might have, and reasons why you would want to work there. This will not only help you stay organized but also be a valuable resource when preparing for interviews.

Remember, by taking the time to research and understand your target companies, you are not only increasing your chances of landing an interview but also ensuring that you're targeting jobs that align with your career goals and personal values. And in a job market where 54% of HR professionals report struggling with finding qualified candidates (SHRM), this targeted approach will help you stand out.

A research journal or spreadsheet allows you to keep all of your research organized and easily accessible.

Here's a detailed example of what this could look like:

Spreadsheet Format

1. **Company Name:** Start with the basics - write down the name of the company.

2. **Industry:** Write down the industry or sector the company operates in.

3. **Company Size:** Note the size of the company. This could be in terms of revenue, number of employees, or both.

4. **Location(s):** Write down the location of the company's headquarters and any other locations that are relevant to your job search.

5. **Mission Statement:** Include the company's mission statement or any guiding principles that are important to its operations.

6. **Products/Services:** Briefly describe the products or services the company offers.

7. **Culture/Values:** Make notes on the company culture and values based on your research.

8. **Job Openings:** List any current job openings that align with your career goals.

9. **Key Facts:** Include any key facts or interesting things you've learned about the company during your research.

10. **Contact Information:** If you have contact information for the company (such as the HR and

recruiting department or a specific staff member or someone in your personal network), include that as well.

11.   **Notes/Impressions:** Leave a section for any general notes or impressions you have about the company. This could also include questions you want to ask during a potential interview.

12.   **Follow-Up:** Finally, include a column where you can note down follow-up actions, like applying for a job or reaching out to a contact.

In addition to keeping you organized, this spreadsheet can be a valuable resource when tailoring your application materials and preparing for interviews. It allows you to quickly reference key information about each company and keeps track of your job search progress.

There are several free resources available online that can assist you in organizing and managing your job search.

Here are a few examples:

1. **Google Sheets Job Search Templates:** Google Sheets offers several free templates for job search tracking. You can easily customize these templates to meet your specific needs.

2. **Microsoft Office Job Search Tracker:** Microsoft offers a free Job search tracker template that helps you keep track of companies you have applied to, contacts, and job descriptions.

3. **JibberJobber:** A job search management tool that lets you track the companies you apply to, the jobs you're interested in, the people you've spoken to, and more.

4. **Huntr:** This tool helps you organize and track job applications, interviews, and contacts. It also features a job search engine that sources opportunities directly from company career sites.

Remember, it's important to choose a system that works best for you to keep your job search organized and efficient.

4.3 Navigating Salary Discussions: Understanding and Negotiating Your Worth

Understanding and negotiating your worth is a skill that many transitioning military personnel might not have had the opportunity to develop. Unlike the military, where pay is standardized based on rank and years of service, the civilian job market can have a wide range of salaries for similar roles, influenced by factors such as location, industry, company size, and the specific set of skills required.

One crucial aspect of your job search is knowing where you want to work. The location will significantly impact the salary you can expect. For instance, a job in New York City is likely to pay more than the same job in a small town in the Midwest, simply due to differences in the cost of living. It's important to understand that, unlike the military, civilian employers often do not provide allowances for the cost of living or housing.

There's no doubt that this chapter will be one you'll find yourself coming back to frequently. As a salary negotiation expert, I've gathered a wealth of advice, practical strategies and proven techniques that can help you confidently navigate these discussions.

Here are some key points we'll be covering in this chapter:

1. **Understanding Salary Ranges and Preparing for Changes in Income:** We'll delve into how to research and understand salary ranges in your target industry and location and what careers look like without housing or cost-of-living allowances.

2. **Assessing Your Worth:** We'll explore how to assess your skills, experiences, and qualifications to determine your worth in the job market.

3. **Negotiating Your Salary:** We'll walk through the steps of negotiating your salary, including how to handle initial salary discussions, make your case, and respond to an offer.

4.   **Considering the Full Compensation Package:** We'll look beyond just salary, considering other aspects of compensation like benefits, bonuses, and perks that can impact your overall compensation package.

By the end of this chapter, you'll have the knowledge and confidence to navigate salary discussions and ensure you're being compensated fairly for your skills and experience. And remember, negotiation is a skill like any other – with knowledge and practice, you can become adept at it. So don't be intimidated by the process. Instead, see it as another aspect of your career transition where you can apply your resilience, determination, and adaptability.

Understanding Salary Ranges

In the civilian sector, salaries can vary significantly depending on various factors such as the industry, job role, company size, and location. Unlike the military, where pay is standardized, civilian salaries are typically based on market rates, which can fluctuate.

One of the first steps in your job search should be researching the salary ranges for your target job role in your desired location. Websites like Glassdoor, Salary.com, Payscale, and LinkedIn Salary Insights provide salary data for a wide range of job titles and locations.

Exercise: Researching and Comparing Cost of Living & Salary Ranges

Objective: Understand the cost of living in your target location(s) and align it with the salary ranges for your desired job role.

Materials Needed: Computer with internet access, a notebook or digital document for notes, and possibly phone for calls or texts.

Steps:

1. Define Your Target Roles and Locations

- List down 3-5 job roles you're interested in.

- Identify 2-3 potential locations (cities or states) where you would like or are willing to work.

2. Research Salary Ranges

- Using websites like Glassdoor, Salary.com, Payscale, and LinkedIn Salary Insights, record the average, low, and high salary ranges for each of your target job roles in each location.

3. Determine Cost of Living

- Visit websites like Numbeo or Cost of Living Calculator to get an estimate of the living costs for each of your target locations.

- If possible, consult with family or friends who live in or near these locations. Ask them about average rent, utility bills, transportation costs, and any other significant expenses.

4. Compare & Analyze

- Match the salary data with the cost of living for each location. Does the average salary support the lifestyle you anticipate in that area? Are there locations where your desired job role tends to pay more but has a similar or lower cost of living?

5. Document Insights

- Note down any patterns or anomalies you observe. For instance, a city might have high salaries, but the cost of living may also be proportionately higher.

6. Bonus Task: Local Legislation

- Research if your target state or city has laws mandating companies to disclose salary ranges in job postings. This can be valuable in understanding if the salaries you see in job listings are mandated and hence likely more accurate.

Example:
Job Role: Data Analyst
Locations: Austin, Texas and San Francisco, California
Average Salary in Austin: $65,000 *Average Salary in San Francisco:* $90,000
Estimated Monthly Rent in Austin: $1,200 for a 1-bedroom apartment. *Estimated Monthly Rent in San Francisco:* $3,500 for a 1-bedroom apartment.
Analysis: While the salary for a Data Analyst is higher in San Francisco, the cost of living, especially rent, is significantly more than in Austin. The salary in Austin might provide a more comfortable lifestyle given the city's lower living costs.

Footnote: Engaging the support of family members or close friends can significantly enrich your research. They can provide fresh perspectives, double-check information, and even uncover opportunities you might have missed.

If you're considering relocating, family or friends based in your prospective locations can provide valuable insights into the local job market and cost of living. Also, they could potentially tap into their local networks for job leads and insider information.

Also, keep in mind that, as of this writing, several states in the U.S., including California, Colorado, Hawaii, and Washington, have passed laws requiring companies to disclose salary ranges in job postings. This can give you a clear idea of what a job might pay before you even apply. It's worth noting that many multi-state and global companies are adopting this practice, even in states where it's not required by law. Always look for this information in job postings, as it can be an excellent starting point for your salary negotiation.

Preparing for Changes in Income

While researching salary ranges, it's essential to consider that civilian employers often do not provide allowances for the cost of living or housing. In the military, Cost of Living Allowances (COLA) and Housing Allowances (HOLA) are part of your compensation, helping to offset higher costs in certain areas.

However, in the civilian sector, these allowances are rarely provided. Your salary needs to cover all your living expenses, including housing. This means that if you're considering a job in a city with a high cost of living, such as New York, Honolulu or San Francisco, you may need a significantly higher salary than in a smaller city or town to maintain the same standard of living.

As you research salary ranges and prepare for potential changes in income, it's crucial to factor in these expenses. Also consider the cost of benefits, like health insurance, that may have been covered by the military but depending on your post-military benefits, could now be an out-of-pocket expense.

Key Takeaways:

- Civilian salaries often don't include allowances like COLA or HOLA.

- High-cost cities may require a considerably higher salary to maintain your current living standards.

- Consider benefits that were once covered by the military but might now be out-of-pocket in the civilian sector.

- Proper research helps set a target salary range and equips you for negotiations.

By doing this research and financial planning, you'll have a clearer picture of the salary you need to aim for in your job search and be better prepared to negotiate your worth.

Assessing Your Worth

In the civilian job market, understanding your worth involves more than just assessing your skills and experiences. It's about knowing the value you bring to a potential employer.

This value can be influenced by a variety of factors, including your technical skills, soft skills, leadership experience, education, and industry certifications.

Here are some concrete steps to help you assess your worth:

1. **Identify Your Skills:** Make a comprehensive list of all your skills - both 'hard' technical skills and 'soft' skills. These can include all of the skills you've been working on in the previous chapters. It could include leadership, problem-solving, communication, adaptability, and others. Many skills you developed in the military, such as discipline, teamwork, and the ability to work under pressure, are highly valued in the civilian sector.

2. **Highlight Your Experiences:** Your experiences, both in terms of the roles you've held and the specific projects or tasks you've completed, can significantly influence your worth. Be sure to consider all relevant experiences, even those that might not seem related to the job you're targeting.

3. **Note Your Achievements:** Document any noteworthy achievements, such as commendations, awards, or significant accomplishments during your military service. These can serve as concrete proof of your skills and abilities.

4. **Education and Certifications:** Your educational background and any industry certifications you hold can also add to your worth. In certain industries, specific certifications can significantly increase your market value.

As American entrepreneur and motivational speaker Jim Rohn once said, "You don't get paid for the hour. You get paid for the value you bring to the hour." Reflect on the unique value you bring and use that to determine your worth.

To help you assess your worth, let's create a worksheet where you can list all of your hard and soft skills, experiences, achievements, and qualifications. This exercise can provide a clear picture of your worth and serve as a useful reference during salary discussions.

Exercise: Skill Articulation and Value Proposition

Building upon your previous exercises, where you practiced using "I" statements and identified your skills and interests, this exercise will delve deeper into differentiating and showcasing hard and soft skills.

Recognizing the distinction is pivotal for your transition, as it will help tailor your approach to different job positions and employers' expectations.

Step 1: Understand the Difference

Hard Skills: These are specific, teachable abilities or knowledge sets that you can quantify. They are often gained through education, training, or experience.

Examples include:

- o Computer programming: specific languages like Python or Java.

- o Equipment operation: operating specific machinery or tools.

- o Certifications: such as CPR certification or project management certification.

Soft Skills: These are interpersonal or people skills. They are harder to quantify and relate to how you work and interact with others.

Examples include:

- o Communication: your ability to convey information effectively.

- o   Leadership: guiding and motivating a team.

- o   Adaptability: adjusting to new situations or challenges.

Step 2: Expand on Your Lists

Skills: Revisit your skills list. Label each skill as hard or soft. For hard skills, be specific about tools, techniques, or methods. For soft skills, think about situations where they played a pivotal role in your military role.

Experiences: Link specific skills to experiences. How did a particular hard or soft skill play into a successful mission or project?

Achievements: Connect achievements to the skills that facilitated them. Did your proficiency in a hard skill lead to an accolade? Or did a soft skill play a pivotal role in a team's success?

Qualifications: Pair any qualification with associated hard skills. For instance, a certification in cybersecurity can be paired with hard skills like "knowledge of intrusion detection systems" or "experience with firewall management."

Step 3: Articulate the Value

For each item, use "I" statements to describe how it contributes value. For instance, "I mastered multiple programming languages, enabling me to create efficient software solutions. In a civilian tech role, I can quickly adapt to diverse tech environments and deliver results."

By the end of this exercise, you'll have an enriched understanding of your hard and soft skills, their interplay in your experiences, and their prospective value in a civilian context. This nuanced understanding is key for crafting impactful resumes, cover letters, and for performing confidently in interviews.

Example: Logistics Specialist
Hard Skills:

Inventory Management: Managed a rotating inventory of over 10,000 equipment items, ensuring a 98% accuracy rate in quarterly audits.

Equipment Maintenance: Proficient in the routine upkeep and troubleshooting of key equipment, reducing equipment downtime by 20%.

Database Usage: Utilized military logistics software to track and allocate resources during deployments.

Soft Skills:

Team Collaboration: Worked in tandem with a team of 12, coordinating efforts to ensure seamless supply chain management during operations.

Time Management: Prioritized tasks during high-pressure situations, ensuring timely delivery of essential supplies to front-line units.

Problem Solving: Addressed and rectified supply shortages during overseas deployments by collaborating with local vendors and improvising with available resources.

Value Articulation: "I utilized specialized logistics software to manage extensive inventories, ensuring efficient allocation of resources. In civilian supply chain roles, I can leverage this expertise to optimize inventory management, reduce costs, and enhance operational efficiency."

Example: Intelligence Officer
Hard Skills:

Data Analysis: Analyzed raw intelligence data to identify patterns, leading to the interception of three major threats.

Cybersecurity Protocols: Implemented and oversaw adherence to advanced cybersecurity measures to protect sensitive data.

Briefing Creation: Developed and presented daily intel briefs to command, utilizing platforms like PowerPoint and specific military software.

Soft Skills:

Leadership: Led a team of 25 analysts, fostering a collaborative environment that increased intel accuracy by 15%.

Strategic Thinking: Planned and executed intel operations based on overarching military strategy, adapting to changing on-ground scenarios.

Communication: Effectively communicated complex intel findings to non-intel personnel, ensuring informed decision-making at higher command levels.

Value Articulation: "I spearheaded intel teams, analyzing vast datasets to identify threats and craft strategies. My leadership and analytical prowess can translate into roles in corporate strategy or data analysis in the civilian sector, ensuring informed business decisions backed by thorough data examination.

Bridging Your Military Skills to Civilian Compensation

Having successfully articulated the value of both your hard and soft skills, the next phase is linking those skills to tangible compensation in the civilian job market.

This is where the importance of effective salary negotiation comes into play. Using the insights from your skills, you can better align your expectations with the market rate for your targeted position and location.

Researching Salary Bands

Hard Skills: These typically provide the most direct comparison points when researching salaries. For instance, an Intelligence Officer's data analysis skills can be directly translated to roles such as a Data Analyst or Business Intelligence Specialist.

Action: Use job boards or salary research websites like Glassdoor, Payscale, or LinkedIn Salary Insights. Input your identified hard skills (like "data analysis" or "inventory management") to find roles that demand these skills. This will give you a salary range for these roles in different industries.

Soft Skills: While they might not be directly linked to specific salary figures, your soft skills will often position you for roles with leadership or strategic responsibilities, which can command higher salaries.

Action: When searching job descriptions, look for keywords that align with your soft skills. Leadership, communication, strategic thinking — roles that emphasize these skills often reflect managerial or supervisory positions. Make note of these roles' median salaries.

Location Adjustments: It's essential to remember that salaries can significantly vary based on location due to factors like cost of living and demand for certain skills.

Action: Once you have a rough salary estimate from your skills research, use tools like the Bureau of Labor Statistics' wage data or NerdWallet's Cost of Living Calculator to adjust these figures based on your target cities or regions.

Industry-Specific Inquiries: Your military role might align more naturally with certain industries. For example, a Logistics Specialist from our earlier example might find a seamless transition into supply chain roles in manufacturing or e-commerce sectors.

Action: Research industry-specific salary benchmarks. Websites like the Occupational Outlook Handbook can provide insights into how compensation varies across industries for similar roles.

Exercise: Salary Research

Tools needed: Writing pad or paper and pen, computer, and internet. Time to research; start with at least one hour.

- List Your Top Hard and Soft Skills: Referring to your earlier exercises, jot down your top 3 hard and soft skills.
- Search for Roles: Use job boards to find positions that align with these skills. For each skill, try to identify at least two job titles.
- Record Salary Ranges: For each identified job title, note the salary range in your targeted industry and location.
- Adjust for Location: Use a cost-of-living calculator to see how these salaries measure up in different cities or regions you're considering.
- Compile & Compare: Organize this information in a table or spreadsheet. This will give you a clear picture of where your skills are most valued and guide your job application and negotiation strategies.

Summing Up: Bridging Skills to Compensation

The journey from identifying your military skills to translating them into the language of civilian compensation is multifaceted, requiring introspection, research, and strategy.

The skills you honed in service, both hard and soft, are invaluable assets that can guide not only the direction of your career transition but also the financial expectations you set. Through diligent research, understanding location-based nuances, and aligning with industry standards, you've laid the groundwork to approach the next phase—negotiation—with confidence and clarity.

As you pivot to the subsequent section on "Negotiating Your Salary," carry forward the insights you've gleaned here. They will serve as your compass, ensuring that as you articulate your worth, you do so with the informed perspective of where your skills stand in the civilian marketplace.

Negotiating Your Salary

Negotiating your salary can often feel intimidating, particularly if you're transitioning from the military, where pay is standardized. However, it's a critical component of ensuring you are adequately compensated for the value you bring to a potential employer. While you've already begun the process of understanding and articulating your value, the journey to ensuring rightful compensation needs a revisit. This reiteration is key to ensuring you're not only recognized but also adequately remunerated for the immense value you offer.

1.   **Do Your Research (Again)**: Though you've previously gathered data, it's vital to refresh your research. Job roles, industry standards, and regional variations change. Utilize platforms like Glassdoor, Payscale, and LinkedIn Salary Insights to reestablish your 'anchor' for salary discussions.

2.   **Reaffirm Your Worth and Perceived Value**: Reflect once more on your comprehensive self-assessment. This recalibration helps in reassessing the salary range that you believe mirrors your true worth in the ever-evolving job market. Remember, it's not just about what you think you're worth but how potential employers perceive the value you can bring.

3.   **Initial Salary Discussions**: Just as before, be prepared for these pivotal conversations. Having an updated range at your fingertips, drawn from your renewed research and self-assessment, ensures you remain on solid ground.

4.   **Receiving an Offer**: Every job offer is unique. Assess it with a fresh lens, considering all aspects of the compensation package beyond just the salary.

5.   **Making Your Case (With Renewed Confidence)**: Should the offer not meet your

93

expectations, rely on your updated research and firsthand experiences to provide a compelling argument for a revision.

6. **Counteroffer**: With an enhanced understanding, frame your counteroffers in a manner that underscores the mutual benefits for both you and the potential employer.

7. **Review the Entire Offer**: As you did previously, examine all facets of the offer. The richness of a job proposal often extends beyond the base salary, encompassing benefits, growth opportunities, and other perks.

8. **Confirming the Offer**: As the final step, always ensure the agreed terms are presented in writing.

Revisiting and refining your approach to salary negotiation is a testament to the dynamic nature of the civilian job market and your evolution within it. Practice mock scenarios with family members or friends. Remember, this negotiation isn't just a financial transaction; it's a dialogue that encapsulates your journey, worth, and unique value proposition. Embrace this process with the knowledge that you don't get what you deserve; you get what you negotiate.

Exercise: Mock Salary Negotiation: Landing a Job at a Big Box Retail Hub

Objective: To simulate the process of salary negotiation for a Logistics professional seeking a career at a retail hub in Houston.

Materials Needed:
Notepad and pen (or digital equivalent)
Printout or digital reference to the research data from previous exercises
Confidence!

Setting the Scene:
Find a quiet space.
Sit comfortably and imagine you're in an interview room at a Houston retail hub. The interviewer, an HR manager or recruiter, is across the table.

Opening Dialogue: Start by expressing gratitude: "Thank you for considering me for the logistics role. I'm truly excited about the prospect of joining (name of company), especially at the Houston hub."

Introduce the Salary Discussion: Smoothly transition: "I'd like to discuss the compensation package for this position. Based on my research and understanding of the role's responsibilities, as well as my skills and experiences, I believe a salary of $120k would be appropriate."

Anticipate a Counteroffer: Imagine the HR manager or recruiter saying, "We typically start our Logistics professionals at around $100k, given the salary band for this role. However, we're willing to go up based on experience and skill set."

Assert Your Value: Respond confidently: "I completely understand your company's standard starting point. However, considering my extensive experience in [specific logistics projects], proven track record in [achievements], and unique skills such as [mention a specific skill], I believe that my value aligns more with the upper end of the salary band. Furthermore, based on market research and the cost of living in Houston, a figure closer to $120k seems more in line with the industry standards for someone with my qualifications."

Show Flexibility (If needed):If you imagine the HR manager or recruiter being hesitant, you could add: "I'm genuinely excited about this role and the opportunity to contribute to (the company's) success. While $120k is my target based on the value I bring, I'm willing to discuss a figure or compensation package that works for both of us. Where is your flexibility?"

Reinforce Your Enthusiasm: End with positivity: "Regardless of the numbers, I truly see myself thriving here and making a significant impact. I'm eager to find a compensation package that reflects both my value and (the company's) vision."

Reflect:After the exercise, take a moment to reflect. How did you feel asserting your value? Were there moments of hesitation? Jot down areas where you felt most confident and areas where you'd like to improve.

Notes: Remember, this exercise is a tool to build confidence and prepare for the actual negotiation. It's crucial to remain adaptable and receptive during the real discussion, as variables can change.

Using Anchors – Non-Negotiables in Salary Negotiation

In the realm of salary negotiations and understanding one's worth, the concept of 'anchoring' becomes crucial. An 'anchor' in this context refers to a reference point that sets the tone for the negotiation. This can be likened to a ship that drops anchor to hold its position, regardless of the current or wind. Similarly, in salary negotiations, your anchors are your non-negotiables that define your perceived value.

Consider a transitioning military person who has decided they need to live in a particular location, say Location X, after their service. This decision becomes an 'anchor' in their job search. It might mean they focus only on jobs in Location X, or remote positions that allow them to live there. The cost of living in Location X, the job market, and the salaries prevalent in that area all contribute to forming another 'anchor' – their expected salary range.

In a similar vein, if they have budgeted for a certain lifestyle or have specific financial goals they want to meet, they have another anchor in the form of the minimum salary they are willing to accept. These anchors are essential in defining their perceived value.

Here's an important thing to note - an anchor isn't just a rigid constraint. It's a reflection of what you value, your needs, and your worth. It's a statement that says, "This is important to me, and I'm worth it." This perspective reinforces the importance of your anchors and gives you the confidence to stand by them during negotiations.

When you approach salary negotiations with clearly defined anchors and an understanding of your perceived value, you're in a stronger position to negotiate a compensation package that aligns with your worth and meets your needs. Remember, the aim is not just to land a job, but to secure a role that respects your value and supports your goals.

Exercise: Defining Your Non-Negotiable Anchors

Having explored the concept of anchoring, it's time to delve into an exercise to identify and define your non-negotiables. This will provide clarity on what truly matters to you, ensuring you remain anchored during negotiations.

1. List Your Priorities: Take a moment to consider what's most important to you in your next role. This could relate to job responsibilities, company culture, location, or compensation. Write down everything that comes to mind.

2. Categorize Them: Group the items on your list into three categories:

- *Essential:* These are your non-negotiables. They are so critical that you wouldn't consider a job offer that doesn't meet these criteria.

- *Preferred:* While important, there's some wiggle room here. You'd like these criteria to be met, but you're willing to compromise if other aspects of the job offer are highly attractive.

- *Nice to Have:* These would be great, but they aren't deal-breakers.

3. Elaborate on Your Essentials: For every item in your Essential category, provide a brief explanation of why it's non-negotiable for you. This will help solidify its importance in your mind and prepare you to articulate its importance to potential employers.

Example:

Essential: A job in Location X.

Reason: My family lives in Location X, and I have committed to being close to them to support their needs.

4. Review the Current Job Market: Based on your non-negotiables, research job openings that fit your criteria. Do they align with your skill set and experience? Are your salary expectations in line with these roles?

5. Test Your Anchors: Imagine you receive a job offer that meets all your essential criteria but falls short in one or two preferred areas. Would you take it? Play out a few scenarios in your head where you might have to make concessions. This will help you truly understand and reaffirm your anchors.

By the end of this exercise, you should have a clear understanding of what you're unwilling to compromise on in your job search. As you proceed with interviews and negotiations, refer back to this list regularly to ensure you remain anchored to your true priorities.

When to Pull Up Your Anchors – When Non-Negotiables Become Negotiable

There are certainly instances where your 'anchors' may need to be reconsidered. The key here is flexibility. While having non-negotiables can guide your career choices, it's essential to remain adaptable and open to opportunities that may require you to pull up your anchors.

Let's consider an example:

Suppose you have a particular interest in working for Company ABC, a leading firm in your desired industry. However, Company X's headquarters are in a city with a high cost of living, which could strain your budget. Or perhaps, the salary range for your desired role at Company ABC is slightly lower than your initial anchor point.

In this scenario, your desire to work for Company ABC might prompt you to reconsider your anchors. Maybe the opportunity to gain experience at such a well-regarded company, the potential for career growth, or benefits like remote working options might outweigh the drawbacks.

Remember, anchors serve as a guideline, but they shouldn't restrict your opportunities. Sometimes, the value of a particular role or company may go beyond the salary or location, such as learning opportunities, work culture, or career progression.

So, while it's important to know your worth and have your anchors, it's equally essential to be open to opportunities. It's about striking a balance between your non-negotiables and the potential value an opportunity might bring. Always weigh the pros and cons before making a decision. Your career journey is a marathon, not a sprint, and sometimes the best opportunities come when we're willing to pull up our anchors and set sail to uncharted waters.

Exercise: Re-Evaluating Your Anchors

In this exercise, we'll revisit the anchors you identified in the previous activity. By evaluating potential scenarios, you'll gain insight into when and why you might consider adjusting your non-negotiables.

1. Reflect on Your Priorities: Revisit the list you created in the previous exercise, especially focusing on your *Essential* category. Ask yourself, under what circumstances would each of these become negotiable?

2. Scenario Building: For each of your *Essential* non-negotiables, imagine a hypothetical job offer that challenges it.

Example:
Essential: A job in Atlanta, Georgia.
Scenario: You receive an offer from your dream company,
but it's in Knoxville, Tennessee.

3. Assess the Trade-offs: For each scenario, list potential
benefits and drawbacks. Would the benefits of the job
outweigh the importance of your anchor?

Example:

Benefits: Prestigious company, higher career growth,
excellent benefits package.
Drawbacks: Away from family, higher cost of living,
unfamiliar city.

4. Rank by Willingness to Adjust: From the scenarios
you've created, rank them based on how likely you'd be to
reconsider your anchor. This will give you a sense of which
non-negotiables might be more flexible than others.

5. Consider External Factors: What external elements
might influence your decision to adjust your anchors? This
could be industry trends, economic factors, personal life
changes, or feedback from mentors and peers.

6. Future Projections: Project yourself five years into the
future. How do you think accepting a role that requires you
to pull up an anchor would impact your career trajectory?
Would it set you on an exciting new path, or would it feel like
a compromise?

7. Decision Journaling: Keep a journal of job offers or
opportunities where you felt the need to pull up your
anchors. Document the decision you made and why. Over
time, this journal can offer valuable insights into your career
decision-making patterns and help in future negotiations.

At the end of this exercise, you'll have a deeper
understanding of the flexibility of your anchors. While it's
essential to have strong guiding principles, it's equally
important to remain adaptable, ensuring you don't miss out
on potentially life-changing opportunities. Remember, every
decision is a learning opportunity, helping shape your career
journey.

Understanding BATNA

BATNA, or Best Alternative to a Negotiated Agreement was first introduced by Roger Fisher and William Ury in their book "Getting to Yes: Negotiating Agreement Without Giving In". Fisher and Ury are renowned experts in negotiation and mediation, and their book has become a classic in the field of negotiation strategy.

The concept of BATNA is essential to any successful negotiation. It is the best outcome that you can achieve if the negotiation fails. By knowing your BATNA, you are in a much stronger position to negotiate effectively.

Understanding your BATNA is a powerful tool. BATNA represents the alternative options available to you if a negotiation doesn't result in a satisfactory outcome. Knowing your BATNA allows you to approach negotiations with confidence and clarity, empowering you to make informed decisions and avoid settling for less than what you deserve.

Here are some tips for determining your BATNA:
1. Brainstorm a list of all of your possible options if the negotiation fails.
2. Evaluate each option and select the best one.
3. Make sure that your BATNA is realistic and achievable.
4. Be prepared to walk away from the negotiation if you cannot reach a deal that is better than your BATNA.

Example: BATNA for the logistics professional who is asking for 120k salary in Houston at Big Box Retail Hub could look like:

▪ Accept a job offer from a competing company for a similar salary. For example, the logistics professional could accept a job offer from another company for a $120k salary.

▪ Start their own logistics consulting business. The logistics professional could use their skills and experience to start their own consulting business, which could potentially generate more income than a traditional job.

- Take a job in a different location. The logistics professional could take a job in a different city, such as Dallas or Austin, where the cost of living is lower, and salaries are higher.

- Negotiate a higher salary. The logistics professional could use their BATNA by pointing out that they have received job offers from other companies for a higher salary.

By identifying and strengthening your BATNA, you position yourself strategically, making it easier to walk away from a deal that doesn't align with your goals.

It is important to note that BATNA is not the same as your "walking point". Your walking point is the lowest salary that you are willing to accept, and you should have that amount figured out by doing the research and the exercises earlier in the chapter.

Let's look at another example. Let's say your salary expectations are in the $115k-120k range. Using BATNA as a guide, and a figure of $100k as your walking point, let's explore two scenarios:

1. Staying Fixed on Your Salary Point of $120,000: Imagine you receive a job offer with a salary of $100,000 which is far too low for you to continue the negotiations, aka your 'walking point'. After turning down the 100k offer, the company counters and increases the salary to $105,000. This is nowhere near your $120k, but above your walking point. How might you get to your anchor of $120,000 based on your research and market value?

o In this situation, you can leverage your BATNA, which could include alternative job offers or opportunities with similar responsibilities that offer salaries closer to your desired $120,000 mark. By staying firm on your salary point you are effectively showcasing your BATNA and you communicate that you are confident in your market value and are willing to explore other opportunities if the current offer doesn't align with your expectations. This assertiveness could lead to further negotiations and increase the chances of reaching a salary closer to your target.

2.   Being Flexible on Your Salary Point Due to Favorable Conditions: On the other hand, you might find that despite the salary offer being $105,000, the job location and cost of living are exceptionally favorable.

o   Your research shows that the cost of living in this location is significantly lower compared to other potential job opportunities. In this scenario, you can use your BATNA to 'pull up your anchor' and make your initial salary anchor of $120,000 negotiable. By recognizing the value of others (things) and demonstrating a willingness to adapt to your expectations, you showcase your open-mindedness and desire to find a mutually beneficial agreement. This flexibility in negotiations can lead to exploring other areas of the compensation package, such as additional benefits to include a potential sign-on bonus, remote work options, or career growth opportunities, ultimately resulting in a well-rounded and satisfactory deal.

In both cases, your BATNA plays a crucial role in guiding the negotiation process. Whether staying fixed on your salary point or being flexible based on other favorable conditions, your ability to utilize your BATNA effectively showcases your preparedness, confidence, and commitment to achieving the best possible outcome for your career transition.

Utilizing Sign-On Bonuses Strategically

A sign-on bonus is a sum of money that an employer offers to prospective employees to incentivize them to accept a job offer. These bonuses can be quite attractive, but it's essential to understand their purpose and how to strategically negotiate for them.

1.   **Why Employers Offer Sign-On Bonuses:** Employers use sign-on bonuses in many ways. They could use it in advertisements to gain more attention and make their job offer more attractive, particularly when they're competing for top talent. If during negotiation if a company cannot meet a candidate's salary expectations due to budget constraints or salary bands, a sign-on bonus can help bridge the gap. Expect sign-on bonuses to be a few hundred dollars to several thousand, depending on the level. Rule of thumb, no more than 10% of a base salary.

2.   **Negotiating a Sign-On Bonus:** If you receive a job offer without a sign-on bonus, and you feel the salary offered isn't competitive, consider negotiating for a sign-on bonus. This could be particularly beneficial if the employer has restrictions on salary but more flexibility with bonuses.

3.   **Understand the Terms:** Often, sign-on bonuses come with strings attached. You may need to commit to staying with the company for a certain period, usually a year or two. If you leave before that time, you may have to pay back the bonus. Always read the terms before accepting a sign-on bonus.

4.   **Tax Implications:** Be aware that sign-on bonuses are generally taxable. They could potentially bump you into a higher tax bracket for the year, which might reduce the net benefit.

5.   **Strategic Use of Sign-On Bonuses:** If negotiated wisely, a sign-on bonus can serve multiple purposes. It can help with out-of-pocket relocation expenses, buffer the financial impact of unemployment, or if the new job's regular paychecks are scheduled differently than your current pay schedule, or help you meet other financial goals faster.

Remember, a sign-on bonus can be a great asset, but it's essential to understand its implications fully. Always consider the terms, your own financial needs, and professional goals before accepting a bonus.

Considering the Full Compensation Package

When evaluating job offers, it's crucial to look beyond the base salary. The full compensation package includes several other components that can significantly affect your overall earnings and job satisfaction. While each company's benefits package can vary, they typically include:

1.   **Health Insurance:** Most companies offer health coverage, including medical, dental, and vision insurance. The coverage details and employee contributions to premiums can vary widely. If you will be receiving healthcare benefits after your service, look into supplemental insurance.

2.   **Retirement and Employee Savings Plans (401k):** Many companies offer a retirement savings plan, like a 401k, often with some level of employer-matching contributions.

3.   **Paid Time Off (PTO):** This includes vacation, sick leave, and personal time. Some companies offer a set number of days, while others have unlimited PTO policies.

4. **Bonuses:** These could be performance-based or discretionary bonuses during your employment. Some companies may also offer sign-on bonuses or relocation assistance.

5. **Equity or Stock Options:** Especially common in startups and tech companies, this gives employees an ownership interest in the company. Other companies may offer a reduced stock buy-in % for employees.

6. **Professional Development Opportunities:** These include access to training and certification courses, tuition reimbursement for further education, and opportunities for career advancement.

7. **Flexible Work Arrangements:** This could include flexible hours, remote work options, or compressed workweek options.

8. **Other Benefits:** These could include life insurance, disability insurance, wellness programs, employee assistance programs, and even unique perks like gym memberships, childcare assistance, subscription legal-aid or pet insurance. For military receiving Tri-Care, some companies may offer a Tri-Care supplement to eligible employees.

It's essential to consider the value of these benefits as part of your total compensation. However, keep in mind that while these benefits add value, most are standardized across the company, and the room for negotiation may be limited. Usually, the most flexible parts of a compensation package are the base salary and the potential for a sign-on bonus.

Remember, a higher salary does not necessarily equate to a better offer if another employer offers a comprehensive benefits package. Take the time to consider your lifestyle, you and or your family's needs, and your financial goals when evaluating job offers. Remember, it's about more than just the paycheck—it's about the overall package.

Relocation Considerations

Let's explore a positive scenario where a company wants to offer you a relocation package to ease the process. As you prepare to transition out of the military, you will likely have a military relocation to your Home of Record.

Corporate relocation packages are often designed to cater to individual needs and can vary based on the level of the job. For instance, a junior employee might receive a different package than a corporate vice president.

These packages can include a range of benefits, such as temporary housing, a rental car, and a stipend for miscellaneous expenses. It's important to note that relocation packages are typically not offered as a cash payment but are administered by a third-party administrator to ensure a smooth and organized process.

As someone who has likely moved countless times during your military career, you may find that corporate relocations offer more flexibility and customization. Unlike military Permanent Change of Station (PCS) moves, which follow standardized guidelines, corporate packages can be tailored to your specific needs and preferences.

Scenario: Imagine you are currently stationed in Alaska, and your home of record is in Indianapolis. You've just received an exciting job offer for a position located in Chicago, Illinois. Now, you have two positive relocation scenarios to consider:

Scenario 1: Leveraging Both Moves: In this scenario, you decide to take advantage of both your military move and the corporate relocation package. You could allow the military to move you from Alaska to your home of record in Indianapolis. This move is often referred to as your "final move" in the military, and it allows you to return to your designated home after your service. Next, you would accept the corporate relocation package from your home of record in Indianapolis to your new job location in Chicago.

This option offers the advantage of using both moves strategically to cover different aspects of your relocation journey. By doing so, you can optimize your benefits and have a smoother transition from the military to your civilian career in Chicago.

Scenario 2: Saving Your Military Move for Later: In this scenario, you decide to save your military move for later and solely rely on the corporate relocation package to move you from Alaska to Chicago. By doing this, you keep your military relocation benefits in reserve and choose to use them at a more opportune time.

Opting for this approach allows you to take advantage of the corporate relocation benefits offered by your new employer while preserving your military move for when it aligns better with your future plans or personal circumstances. This strategic decision gives you greater flexibility and control over your relocation process and can provide additional support as you settle into your new civilian career in Chicago.

One significant difference between military PCS moves and corporate relocations is the ability to strategically use your military move in conjunction with a company's relocation package. You may have the option to save your last military move for later, allowing you to take advantage of the benefits it offers at a time that suits your circumstances best.

Ultimately, the choice between the two scenarios depends on your individual preferences, financial considerations, and long-term career goals. By carefully evaluating the benefits and potential implications of each option, you can make an informed decision that best suits your needs and sets you up for a successful transition to your new job in Chicago.

Disclaimer: It's important to note that military relocation benefits and policies may vary and are subject to change. Before making any decisions regarding your relocation, it is highly recommended to reach out to the appropriate personnel at your military base or installation to gather up-to-date information about available benefits and entitlements.

The military transition process can involve complex considerations, and understanding the latest guidelines for relocation assistance can significantly impact your decision-making. Consulting with the relocation office or transition assistance program at your base will ensure that you receive accurate and tailored guidance based on your specific circumstances.

By staying informed about the most current information, you can make well-informed choices about how to best utilize your military relocation benefits in conjunction with any corporate relocation package offered by your future employer. Remember, being proactive and seeking guidance from reliable sources will help you navigate your transition smoothly and maximize the support available to you during this exciting new chapter of your career.

PAT YOURSELF ON THE BACK!
12-MONTH COUNTDOWN REVIEW

You've made an incredible journey so far! You've come a long way from where you started, laying a firm foundation for your transition into the civilian job market.

You've identified your transferable skills and interests, considered various career paths, networked like a pro, created one or several resumes and strategized your approach. And now, you're deep in the trenches of the final year. You should be proud of yourself for all that you've accomplished so far!

At this point, it's normal to feel a bit overwhelmed. The last year is often intense, filled with the hustle of job applications, interviews, and, most importantly, salary negotiations. But remember, you're well-equipped to handle this! You've done your homework, and you've got a wealth of knowledge at your disposal, not to mention the tremendous skills and experience you bring from your military service.

Negotiating a fair salary is one of the most critical steps in this process. The ability to negotiate successfully for what you're worth is empowering and sets a precedent for your future career growth. So, keep that energy up! The work you're doing now is setting the stage for your success in the civilian sector. You're not just looking for a job; you're working towards a fulfilling and rewarding career.

And remember, you're not alone in this. Whether it's mentors, peers, family, or friends, people are cheering you on. With every step you take and every challenge you overcome, you're proving your resilience and dedication.

Now, let's take a deep breath, reset, and prepare to tackle the next exciting phase. This last year is where all your careful preparation pays off, where you make your move, assert your worth and land that dream job. So, let's get ready for it! Remember, you're strong, you're capable, and you've got this!

What The Next 12 Months May Look Like:
Having a defined timeline can make the process less overwhelming and provide clear milestones to work towards. Let's break down what your 12-month countdown could look like:

Months 1-3: Foundations and Preparation Start this crucial year by reviewing Chapters 1-4. You've done considerable work already, and now it's time to ensure all your tools are ready. Have your civilian and federal resume templates prepared, tailored to highlight your skills and experiences most relevant to your desired roles.

Identify your preferred location(s) post-service. Research the cost of living, job opportunities, and the professional landscape in these areas. Have a clear idea of the types of jobs you're interested in and conduct preliminary research on salary ranges in your chosen industry and location. Use these figures to start thinking about your salary expectations.

Months 4-6: Networking and Visibility Now that you have your foundations set, it's time to increase your visibility and strengthen your networks. This means actively participating in online forums, LinkedIn groups, and industry events related to your field.

Use this time to connect with individuals who can offer insights into your chosen career and advice on your transition. Set up informational interviews, ask meaningful questions, and learn from their experiences.

Don't shy away from expressing your career aspirations and your timeline for transition. The more people are aware of your plans, the more they can offer timely opportunities, advice, or connections.

Also, start strategically engaging with companies you're interested in. Follow their activity on LinkedIn, comment meaningfully on their posts, and become a familiar name.

Months 6-9: Training Completion and Skillbridge Programs As you're ramping up your networking activities, ensure you're also on track to complete any remaining training or certification courses. Having these credentials in hand will boost your confidence and enhance your marketability as you approach your job search.

This is also the time to finalize your participation in any DoD Skillbridge programs. These programs offer valuable industry experience and can often lead to job opportunities post-service.

Months 9-12: Actively Engaging with Potential Employers, the final stretch of your transition should be dedicated to actively engaging with potential employers. By now, your network should be well-established, and your industry visibility should be on the rise.

About 90 days before your transition, start applying for jobs in earnest. Remember, the job application process can take time, so starting early ensures you're not rushed.

Attend interviews, meet hiring managers, and begin negotiating job offers. This period is where your previous work pays off and your next chapter begins.

Remember, this timeline is a guide, and everyone's transition process will look slightly different. Stay adaptable, remain focused on your goals, and continue moving forward. You're on the path to an exciting new career. You've got this!

CHAPTER 5:
THE FINAL PUSH - 6 MONTHS OUT

5.1 Navigating the Final Stages of Your Military to Civilian Career Transition

As you enter the final stages of your military-to-civilian career transition, it's important to recognize that these months can be an emotional rollercoaster. Amidst all the logistical details of out-processing, attending medical appointments, and finalizing your transition paperwork, there's an underlying psychological transition that you're navigating as well.

Leaving the military means leaving behind a structured environment where your role and responsibilities are clear-cut. Stepping into the civilian world can feel like stepping into a completely different universe, where the rules are less explicit and the path less defined.

In this uncertain period, resilience and self-care become even more vital. Your mental health is as important as your physical health, and nurturing it is crucial for a successful transition.

Here are a few points to consider:

1. Embrace the Change: Acknowledge that this transition is a significant life change and it's okay to have mixed feelings. "When we finally arrive in the private sector, the lack of a deeply personal connection is painful and sometimes debilitating. The challenges we face after transition are exacerbated when immersed in this foreign environment. The same struggles associated with the communication and cultural gap persist, and new ones emerge. Transition is a challenge. And my challenge to you is that when you succeed, turn around and help another military veteran. Bring that veteran into your 'perimeter.' Never leave a fallen comrade!" - John Buckley, Koch Military Relations Manager (from the article "Transition Is a Challenge: Tips for Success in the Private Sector")

2. Cultivate Resilience: Resilience, or the ability to bounce back from adversity, is crucial during this transition. Dr. Kenneth Ginsburg, human development expert, describes resilience as a "natural counterweight" to life's stressors. He suggests the 7 C's to resilience: competence, confidence, connection, character, contribution, coping, and control.

3. Competence – Your ability to handle stressful situations effectively is crucial during this phase of your life. It involves acquiring and practicing essential skills to tackle challenges head-on. Look for resources that offer stress-reduction and social skills training, creating a supportive environment where you can hone these abilities and build your confidence in dealing with various situations.

4. Confidence – Believing in your capabilities is the foundation of self-assurance. As you demonstrate your competence in real-life scenarios, your confidence grows. Seek resources that identify your strengths and acknowledge your accomplishments, empowering you to face challenges with renewed self-belief.

5. Connection – Cultivating strong ties with friends, family, and community groups enhances your sense of security and belonging. These connections anchor your values and reduce the inclination to pursue destructive alternatives. Find resources that emphasize the importance of building meaningful relationships as you embark on your civilian career.

6. Character – Your sense of self-worth and confidence play a vital role in your transition. Being in touch with your values and sticking to them enables you to demonstrate empathy and make wise choices aligned with your principles. Seek resources that focus on enhancing your self-esteem, empathy, and care for others during this transformative period.

7. Contribution – Discovering the power of personal contribution to the world is a profound lesson during your transition. Recognize how your actions can create a positive impact, fostering a sense of purpose and connection. Explore resources that help you explore ways to contribute and make a meaningful difference in your new community.

8. Coping – Equipping yourself with a diverse set of coping skills, such as stress reduction and social techniques, equips you to handle life's challenges effectively. Look for resources that provide practical tools and techniques to manage everyday stresses during your transition.

9. Control – Realizing that you have control over your decisions and actions empowers you to navigate challenges confidently. Making choices aligned with your values and goals enables you to rebound from adversities. Seek resources that emphasize taking charge of your journey, making informed decisions, and embracing opportunities as you transition into civilian life.

10. Practice Self-Care: Remember to take care of your physical health as well. Regular exercise, a balanced diet, and adequate sleep can significantly impact your mental well-being. Take breaks when needed and engage in activities that you enjoy.

11. Seek Support: Don't try to navigate this transition alone. Connect with veterans' support groups, family, friends, mentors, and mental health professionals who can provide emotional support and advice.

Military OneSource, a DOD-funded program, offers free, confidential mental health resources for veterans and their families. Additionally, the Real Warriors Campaign, funded by the Defense Health Agency's Psychological Health Center of Excellence, also offers mental health resources and a peer support community.

There are several other professionals who offer services specific to the emotional aspects of transition, and you might find solace and guidance from these resources. As of this writing, Dr. Shauna Springer, often referred to as "Doc

Springer" within the military community, stands out as a leading expert on trauma, military transition, and close relationships.

Josh Goldberg holds the mantle of Executive Director at the Boulder Crest Foundation, a dedicated organization concentrating on the wellness of combat veterans and their families.

Similarly, Dr. Abby Cobey, a psychologist, directs her profound expertise in traumatic stress towards assisting veterans, particularly in collaboration with the PTSD division of the VA.

Dr. Bridget C. Cantrell, with her vast experience, offers specialized mental health counseling to veterans and their family members, ensuring they receive the support they deserve.

Dwayne Paro, drawing from his own military experiences, serves as a life transition coach. He has crafted his skills to aid veterans as they navigate the intricate shift from military to civilian life.

In a similar spirit, Duane K. L. France, a retired military combat veteran, has pledged his post-military journey to provide mental health counseling for his fellow veterans. This commitment speaks volumes of his dedication to the cause.

Rounding off this list is Dr. Shannon Curry, a Clinical Psychologist and the Director of the Curry Psychology Group based in Newport Beach, California. She specializes in evidence-based treatments tailored for the diverse challenges veterans face.

Each of these professionals, equipped with their unique expertise and experiences, serves as a beacon for veterans, guiding them through the complexities and triumphs of post-military life. For those interested, a simple online search will provide more information about each of these remarkable individuals and their work, or you may find other resources through your Family Readiness office at your current military base.

The upcoming transition may seem daunting but remember that you're not alone. Many have navigated this journey before you and have emerged successful on the other side. As you step into this new phase, take with you the strength, discipline, and resilience that your military service has instilled in you. You are well-equipped to face any challenges that lie ahead.

DoD Skillbridge Programs or Registered Apprenticeship Programs (RAP)

This is also the time to finalize your participation in any DoD Skillbridge or Apprenticeship programs. These programs offer valuable industry experience and can often lead to job opportunities post-service. If you've been accepted into a program, this is when you'd typically start. Your participation in the Skillbridge program will likely inform your job applications and interviews during this period.

As you navigate these final six months, remember the foundation you've built and the work you've invested in this transition process. You're well-prepared to tackle these final stages and make a successful transition into your civilian career.

Applying for Jobs: Where to Look and How to Stand Out

By now, you should have a clear vision of the career path you wish to pursue, and your resume(s) and LinkedIn profile should be fine-tuned for the jobs you're targeting. It's time to start applying for jobs but with a strategic and focused approach.

Remember that job boards and company websites aren't the only places to look for job opportunities. Your network, which you've been cultivating, can also be a great source of leads.

Reach out to your contacts, let them know you're entering the job market, and inquire if they know of any suitable opportunities.

Advice and Steps

Understand the Job Description: Ensure you understand the requirements of the job and how your skills align.

1. Tailor Your Application: Each job application - resume, cover letter, etc., should be specifically tailored to the job you're applying for.
2. Use Relevant Keywords: Pay attention to the keywords in the job description and make sure to include these in your application where relevant.
3. Use Online Tools: Consider using online tools like SkillSyncer to match your resume with the job description for a higher chance of getting through applicant tracking systems. Also, using an AI plug in to help tailor your resume to the specific job description could be useful.

Worksheets and Tools:

- There are new productivity tools released weekly, do an internet search and find one you like, or create an XLS worksheet in Excel

Popular Job Boards (Non-Cleared Jobs):
- Indeed
- LinkedIn Jobs
- Glassdoor
- Monster
- CareerBuilder
- TealHQ

Popular Job Boards (Cleared Jobs):
- ClearanceJobs
- Dice
- Cleared Connections
- IntelligenceCareers
- TechExpo

Again, please remember to verify the legitimacy of any platform before submitting your personal information.

Personal Cyber Security

In today's digital age, safeguarding your personal information has never been more critical. When it comes to job applications, it's essential to understand the difference between what you should include on your resume and what you should only provide during the official application process. Your resume should be limited to professional details such as your skills, experiences, and achievements. The only personal information that should be on your resume includes your name, city, state, zip code, professional email address, and contact phone number.

Do not include sensitive personal information like your social security number, date of birth, or full home address on your resume. This information should only be provided during the formal application process on a trusted, secure platform. Remember, resumes often get distributed widely during the hiring process, and you wouldn't want such sensitive information falling into the wrong hands.

For those with security clearances, such as secret, top secret, TS-SCI, and CI or FS poly, the stakes are even higher. Malicious actors can target individuals with access to sensitive information, and any exposure to personal information increases the risk of being targeted. It's also worth noting that while job boards can be a valuable tool for your job search, they can also be a vector for these malicious actors. Some nefarious individuals may even buy a seat on a job board with the sole intention of extracting sensitive information. As such, always verify the legitimacy of the job board, the job posting, and the company before submitting your application or any personal details. Practice good cyber hygiene by securing your personal information and being mindful of where and how you share it.

References: Regarding references, it's considered best practice to leave them off your civilian corporate resume.

Your resume should be a document that promotes your skills, experiences, and achievements, and it's not typically the right place for listing references. More importantly, providing references upfront could risk those individuals being contacted prematurely or without your knowledge, which is not ideal. Most employers understand this and will ask for references when they're needed, typically during the interview process or pre-offer stage. When that time comes, it's good to have a separate document ready, with the names, titles, and contact information of your references, along with a brief note about your relationship to each reference (i.e., former manager, coworker, etc.). Always remember to seek permission from your references before providing their contact information to a potential employer.

Also of value, make sure your references are aware that you are providing their names, email, or cell number. Recommend that you ask before you list them or provide them to a potential employer.

To stand out, remember to customize your application materials (resume, cover letter) to each specific job. Emphasize your skills and experiences that align with the job description. Don't forget that as a military veteran, you possess unique experiences and skills that set you apart.

Language is Important – Using Civilian-Speak

Several companies and organizations offer services to help translate Military Occupational Specialties (MOS) into civilian language. Here are a few:

1.   **CareerOneStop.org:** Sponsored by the U.S. Department of Labor, CareerOneStop offers a Military to Civilian Occupation Translator tool that helps identify civilian careers that match the skills gained in a military role.

2.   **O*Net OnLine:** This tool provides detailed descriptions of the world of work for use by job seekers. Its Military Crosswalk Search allows you to enter a code or

title from the Military Occupational Classification (MOC) and get a list of civilian occupations that have similar skills or experience.

3.   **Hire Heroes USA:** This organization offers a resume builder tool that helps to translate military experience into skills sought by civilian employers.

4.   **Military.com:** In partnership with Monster, Military.com offers the Military Skills Translator tool that helps translate military experience into a powerful civilian resume.

Many Mil2Civ, or Veteran-supportive companies are integrating MOS translators into their own applicant tracking systems, and external career sites. Do a quick search to find the latest and give them a test!

5.2 Interviewing Like a Civilian: Tips and Techniques

The interview process in the civilian world can significantly differ from what you're used to in the military. Here are some tips to navigate this process successfully:

1. Prepare: Conduct thorough research on the company and its role. Familiarize yourself with the job description and be prepared to discuss how your skills and experiences make you a good fit.

2. Practice: Conduct mock interviews to get comfortable answering common interview questions. Practice translating your military experiences into civilian terms.

3. Be Professional: Dress appropriately, arrive on time, and maintain a respectful demeanor.

4. Show Enthusiasm: Demonstrate your passion for the role and the company.

5. Ask Questions: Remember, an interview is also for you to learn more about the company and role. Prepare thoughtful questions about the role, the team, and the company culture.

Transitioning from military to civilian life is a significant change, and it's completely normal to feel a mixture of excitement, anxiety, and uncertainty. When it comes to job interviews, nerves can often kick into high gear. Remember, it's okay to be nervous. It's a sign that you care about the opportunity in front of you.

However, it's essential not to let those nerves overpower your authentic self. Interviewers want to see who you are, beyond the qualifications listed on your resume. They're looking for a glimpse of your personality, your values, and your approach to work. When you show up authentically, it helps the interviewers see how you might fit into the team and the company culture.

So, take a deep breath, remember the skills and experiences you bring to the table, and just be yourself. You've navigated military life, and you're equipped with resilience, discipline, and a wealth of unique experiences. Bring those same strengths to your job interviews. After all, authenticity resonates, and it's your most powerful asset.

5.3 Evaluating Job Offers and Negotiating Salaries

Receiving a job offer is exhilarating but resist the urge to accept it immediately. It's standard practice to take at least 24 business hours to consider an offer thoroughly. This waiting period isn't just about taking the time to celebrate your success, it's about ensuring the decision you make is in your best interest.

In your evaluation, consider the full compensation package. This includes not only the salary but also benefits such as health insurance, retirement contributions, and any bonuses or stock options. Furthermore, aspects like work-life balance, company culture, and growth opportunities should also factor into your decision.

Remember, a job isn't just about the paycheck - it's about your overall satisfaction and future prospects in the role. Take the time you need to weigh all these elements before you give your response. And, if needed, don't hesitate to negotiate for better terms that accurately reflect your skills and experiences.

When negotiating salaries, recall the research you did earlier in this transition process. Understand the market rate for the role in your chosen location, and don't hesitate to ask for a salary that aligns with your skills, experiences, and the cost of living in the job location.

If the offered salary isn't as high as you expected, consider negotiating for a sign-on bonus, which can help bridge the gap.

As we've outlined in the book, gaining strength in learning the process of negotiation is a lifelong skill. It can give you great strength and confidence.

When evaluating job offers and negotiating salaries, consider all the aspects we discussed in Chapter 4.

1.   Understand the salary range for your target role in your desired location. Use online resources like the Bureau of Labor Statistics, Glassdoor, and PayScale to research industry-standard salaries. Knowing this range will give you a benchmark to evaluate the salary offered in your job offer.

2.   Assess your worth. Draw on your skills, experiences, achievements, and qualifications to determine what value you bring to the potential employer. This will help you understand what a fair salary would look like for you specifically, not just what the industry standard might be.

3.   Next, consider the full compensation package. Remember, the salary is only one part of your total compensation. Other aspects to consider include health benefits, retirement plans, paid time off, sign-on bonuses, relocation assistance, and other perks.

4.   When you receive a job offer, take some time to evaluate it thoroughly against your research and your personal needs and circumstances. Resist the urge to say "yes" or "no" immediately. Shoot for a minimum of 24 (working) hours.

If the salary offer falls short of your expectations or the industry standard for your role, be prepared to negotiate. This is where your research, self-assessment, and understanding of your 'anchors' or non-negotiables come in handy. Set your walking point and create your BATNA. Craft a well-reasoned argument for why you deserve a higher salary, referencing the data you've gathered and the value you bring.

Remember, negotiation is a normal part of the job offer process, and your potential employer will expect it. It's not about winning but arriving at fair compensation for your skills and experiences. Stay professional and positive throughout the process and remember that your goal is to build a constructive relationship with your potential employer.

As you prepare for salary negotiation, consider your 'anchors', or the factors that are non-negotiable for you. Your anchors could be a specific salary range you need to maintain your current lifestyle, a certain location you wish to work in, or perhaps specific benefits that are important to you, like health insurance or retirement contributions.

Knowing your anchors in advance will give you a solid footing in negotiations. If you've determined that a certain salary range is your anchor because it's what you need to sustain your current lifestyle, you can approach negotiations with a clear bottom line. Similarly, if the location is your anchor, you'll know to focus on jobs in that specific area or jobs that offer remote work.

However, also be aware that you may need to pull up an 'anchor' at times. For instance, you may be anchored in a certain location, but an incredible job opportunity comes up elsewhere. In such a scenario, you may need to weigh the value of that opportunity against your original anchor. Is it worth relocating for? Are there other aspects of the job offer, like increased salary or career advancement opportunities, that might make you reconsider your original anchor?

Having a clear understanding of your anchors can help you navigate job offers and salary negotiations more effectively. It gives you a strong sense of your boundaries, but also the flexibility to recognize when an opportunity might be worth stretching those boundaries.

This mix of firmness and adaptability will serve you well as you evaluate job offers and negotiate salaries. Remember, the ultimate goal is a fair compensation package that aligns with your skills, experiences, and personal needs.

CHAPTER 6:
THE HOME STRETCH - YOUR LAST MONTH IN THE MILITARY

Let's take a moment to reflect on where we've been so far:

In **Chapter 1**, we discussed what to expect from your journey ahead. We navigated through the civilian job landscape, addressed the common challenges you might face, and underscored the importance of starting early.

Chapter 2 was all about laying the groundwork three years out from your transition. We touched upon identifying your transferable skills, exploring career paths, expanding your network, and enhancing your professional credentials with relevant civilian certifications and training.

In **Chapter 3**, we began to strategize. Two years out from your transition, we started building strong resumes and LinkedIn profiles, expanding strategic networks, and setting clear career goals.

Chapter 4 saw us gaining momentum as we approached the one-year mark. We prepared for the job search, honing our cover letters and interview skills, and we learned to target specific companies and industries. We also dove deep into understanding salary negotiations and the significance of considering the full compensation package.

Chapter 5 marked the final push, with the last six months devoted to actively applying for jobs, acing interviews, and evaluating job offers while fine-tuning our salary negotiation skills.

Now, let's embark on **Chapter 6**, the home stretch of your journey:

6.1 Managing the Emotional Aspects of Transition

Transitioning from the military to a civilian career isn't just a professional change; it's a profound personal journey that can stir a wide range of emotions. It's normal to feel a mix of excitement and anxiety, anticipation, and fear. You're leaving a world you know well for uncharted territory. And while change brings opportunity, it also comes with uncertainty.

One of the most significant challenges you may face is the shift in identity. For years, your role in the military has been a core part of who you are. It's shaped your daily routines, your community, and how you see yourself. Transitioning to a civilian career may feel like you're losing a part of your identity.

Recognize that it's okay to grieve for this. Grieving is not a sign of weakness; it's a sign of respect for what has been a significant part of your life. Acknowledging this loss is an essential step in the transition process.

You may also grapple with feelings of isolation. The camaraderie and brotherhood found in the military are unique, and the civilian world can sometimes feel unfamiliar and lonely in comparison. Remember, it's okay to miss your former colleagues. It's okay to feel out of place.

The key to navigating these emotional ups and downs is to approach them with kindness and patience. Be patient with yourself as you navigate this change. Understand that it's a process, and there will be moments of both progress and setback. And that's perfectly okay.

Stress and anxiety can also be common during this time. Techniques such as mindfulness, exercise, and good sleep hygiene can be beneficial. Remember, self-care isn't a luxury; it's a necessity.

Don't hesitate to seek support, either. Reach out to your fellow veterans who have gone through the same journey. Connect with the professional resources provided earlier, like the mental health and wellness professionals. They can provide strategies to help you manage the emotional aspects of transition.

Lastly, it's essential to remember that despite the challenges and uncertainties, you are not alone on this journey. Countless veterans have walked this path before you and successfully transitioned to fulfilling civilian careers. Their stories serve not only as inspiration but also as proof that it can be done. So, as you face the emotional aspects of your transition, remember to give yourself the grace and space to feel, process, and grow.

Mental health is a critical component of our overall wellbeing, and it's particularly significant during periods of significant change and stress, like the military-to-civilian transition. Unfortunately, some veterans face substantial mental health challenges during this period, and in extreme cases, some may even contemplate suicide.

It's an issue that's far too prevalent and devastatingly real within the military and veteran communities. In facing this, it's important to remember that there is help available, and reaching out is not a sign of weakness, but rather a sign of strength.

If you or someone you know is in crisis, the Veterans Crisis Line offers confidential support 24 hours a day, seven days a week, 365 days a year.

Veterans and their loved ones can call 1-800-273-8255 and Press 1, send a text message to 838255, or chat online at VeteransCrisisLine.net/Chat to receive free, confidential support, even if they are not registered with the VA or enrolled in VA healthcare.

Remember, it's okay to ask for help, and there are numerous resources available specifically designed to support you through these challenges. You don't have to go through this alone.

Here are some of the mental health resources available to veterans:

1. **Veterans Crisis Line** - Offers free, confidential support to veterans in crisis, as well as their family and friends. Available 24/7, 365 days a year.

- Phone: 1-800-273-8255, Press 1
- Text: 838255
- Online Chat: VeteransCrisisLine.net/Chat

2. **National Resource Directory** - An online directory for veterans, service members, and their families to connect with resources and services at the national, state, and local levels.

3. **VA Mental Health Services** - Offers a wide range of mental health services to veterans, including therapy, counseling, and medication.

4. **Give an Hour** - Provides free mental health services to U.S. military personnel and families affected by the most recent conflicts in the middle east, including Iraq and Afghanistan.

5. **Real Warriors** - A multimedia public awareness campaign designed to encourage help-seeking behavior among service members, veterans, and their families coping with invisible wounds.

6. **Coaching Into Care** - A national telephone service of the VA which aims to educate, support, and empower family members and friends who are seeking care or services for a Veteran.

7. **The Soldiers Project** - Offers free, confidential, unlimited mental health services to any active-duty service member or veteran who has served since September 11, 2001.

8. **The Headstrong Project** - Provides cost-free, bureaucracy-free, stigma-free, confidential, and effective mental health treatment for post-9/11 veterans and their families.

This is not an exhaustive list, and I encourage you to reach out to your local VA, your base Military and Family Support Center, your command leadership, or other community organizations to explore more options.

You're not alone, and help is available.

6.2 Final Preparations: Relocation Military vs. Corporate Packages

In this section, we'll help you tie up loose ends and prepare for what's next. We'll discuss practical considerations like mortgages, housing, and relocation, and provide a checklist of final tasks to ensure a smooth transition. This is your time to put all your planning and preparation into action.

Relocating for a job can be a significant decision with its own set of advantages and disadvantages. As a transitioning military member, you'll be facing a unique situation. The military provides a relocation package back to your Home of Record (HOR) or Place Entering Active Duty (PLEAD). On the other hand, some civilian companies might offer relocation packages as part of their job offers, while others might not.

What are the Benefits of the Company's Relocation Offer?

Covered Costs: The most apparent benefit is that a relocation package could cover your moving costs. This could include expenses like moving services, packing, and unpacking services, transportation, and temporary housing.

Eases Transition: A relocation package can ease the financial burden and stress of moving, allowing you to focus more on settling into your new role.

Indication of Company's Investment: A company willing to offer a relocation package might be more committed to your long-term success within the organization.

Disadvantages of the Company's Relocation Offers

Limited to Specific Location: If you're not sure about the location of your new job or if you were planning on moving somewhere else closer to family or friends, a company's relocation package might restrict your options.

Tax Implications: Depending on your situation and the specifics of the package, there may be tax implications.

Ties to the Company: Some companies may require you to stay with the company for a certain period after your move or else you would have to repay the relocation costs. This might limit your flexibility if the job isn't a good fit.

Benefits of the Military's Relocation Package

Flexibility: The military's relocation package to your HOR or PLEAD offers flexibility. You can choose where you want to go and when within a certain timeframe.

Familiarity: Moving back to your HOR or PLEAD might mean returning to a familiar environment, which could ease your transition.

Disadvantages of the Military's Relocation Package

Lack of Coverage for Further Moves: If you use your military relocation package and then decide to move again for a job, you'll likely be covering those costs out of pocket.

Potential Disconnect with Career Opportunities: Your HOR or PLEAD may not align with the best locations for your career aspirations.

When considering relocation packages, it's essential to consider your situation, career goals, and financial implications. A financial advisor or a professional familiar with military transitions can provide guidance tailored to your unique circumstances.

Is your Military Relocation an Asset for Negotiations?

Yes, absolutely! Having a military relocation package available can certainly be an asset during job negotiations. Here's why:

- Negotiating Power: Knowing you have a military relocation package at your disposal means you're not solely dependent on a company's relocation offer. This situation can strengthen your negotiation position. You might choose to negotiate for other elements of your compensation package instead, such as a higher salary or additional benefits.

- Flexibility: Having the military relocation package allows you to be more flexible with your relocation timeline. You can coordinate the best time to move that aligns with your career transition and personal life.

- Leverage: If a potential employer is on the fence about offering relocation assistance, knowing that you have a portion of the relocation expenses covered may encourage them to extend the job offer or enhance other elements of your compensation package.

However, it's crucial to ensure the details of your military relocation package are clear, and that you understand any associated obligations. Also, be sure to communicate clearly about this during your negotiations with potential employers to avoid any misunderstandings.

Home of Record (HOR) and Place of Entry on Active Duty (PLEAD) are official military terms that have specific legal definitions and implications in the military.

- The Home of Record is the place you lived when you entered the military. It is often used to determine certain benefits, such as travel allowances during a permanent change of station (PCS) or separation. The HOR can impact your entitlements related to moving costs, so it's important to understand how it's defined.

- The Place of Entry on Active Duty is the location where you officially entered the military. This could be where you enlisted or were commissioned, or where you started your initial active-duty training.

Generally, changing the HOR or PLEAD is not typically allowed unless there was an error when it was initially recorded or a change in military status, like reenlisting after a break in service.

For military members considering a different location at the end of their service, it's not so much about changing the HOR or PLEAD but planning their final move strategically. It's important to understand the specifics of what the military covers in a final move.

Upon separation or retirement, military personnel are usually entitled to a final move. The military covers the costs of moving to the Home of Record or the Place of Entry on Active Duty, or a location of equal or lesser distance.

However, if a service member decides to move to a location further than their HOR or PLEAD, they may have to pay the difference in cost. The specifics can vary, and there may be additional regulations or allowances depending on the branch of service and individual circumstances.

Always verify the current regulations with your transportation office or legal counsel, as policies may change. Additionally, consider seeking advice from a financial advisor or relocation expert familiar with military transitions to understand all potential costs and benefits associated with your move.

6.3 Securing Your Future - Home Ownership and Job Offers – Why Timing is Critical

One of the most significant steps in the process of transitioning from military to civilian life is securing your future in your new location, and this often involves purchasing a new home. However, it's crucial to understand the process and how your job status - especially a written job offer - can play a critical role.

Securing a mortgage for a new home relies heavily on your income stability, which lenders typically measure through steady employment. Brandi Brickler, a veteran, former military spouse, and mortgage expert, emphasizes that "Lenders need to know that you have a reliable source of income to meet mortgage repayments. A written job offer can be critical evidence of this income, particularly for those transitioning from the military to civilian careers."

When you receive a written job offer, it not only signals your new employer's commitment but also acts as a substantial document indicating your future income. "A job offer letter detailing your start date and salary can often satisfy the requirements of mortgage lenders," says Brickler. "This is particularly valuable for military personnel transitioning to civilian careers, who might not have pay stubs or tax returns from a civilian job when applying for a home loan."

However, it's important to note that lenders may have different requirements and policies, and some might require additional proof of employment, such as a few months' worth of pay stubs. This can potentially delay the home-buying process until after you start your new job. Therefore, early communication with your lender about your job transition is key.

If you're transitioning from the military and planning to buy a home, consider these steps:

1. Talk to a mortgage expert who works with Veterans: Early in your transition process, consult with a mortgage expert who serves the state where you will buy your home. They can guide you through the mortgage process and advise you on what documents you'll need for your home loan application. Timing is critical.

2. Secure a written job offer: As you negotiate job offers, ensure you get the offer, including your start date and compensation, in writing. This document can be critical when applying for a mortgage.

3. Communicate with your lender: Let your lender know about your transition from military to civilian employment and discuss the possibility of using a job offer as proof of income. They can guide you on their specific policies and what further documentation may be needed.

The timing of this phase can indeed be tricky and requires careful planning. Here are a few factors to consider:

- Home loan pre-approval: It's usually a good idea to get pre-approved for a home loan before you start house hunting. Pre-approval involves a more detailed check of your finances and can give you a clear idea of how much you might be able to borrow. However, it's important to note that pre-approval often requires proof of income, which you might not have until you receive a written job offer.

- Job offers and income verification: Mortgage lenders typically need to verify your income before finalizing your home loan. This often requires a job offer letter and a few months of pay stubs. If you're transitioning from the military and don't yet have a civilian job, you might need to discuss with your lender how to use a job offer as proof of income. In some cases, lenders may require you to start the job and provide pay stubs before finalizing the loan.

- Home search and purchase timeline: Once you have a job offer and loan pre-approval, you can start house hunting. After you make an offer on a house and the seller accepts, it typically takes about 30 to 45 days to close on the home. During this time, your lender will finalize your loan.

- Relocation and job start dates: The timing of your relocation and job start date can also be a challenge. Ideally, you'd want to close on your new home and move in before starting your new job. However, this might not always be possible, depending on your job and relocation timelines.

Given these factors, the timing of this phase can be a delicate balancing act. It's essential to communicate effectively with your potential employer, your lender, and your real estate agent to align these timelines as closely as possible. Planning, staying organized, and being ready to adjust your plans as needed can help make this process go more smoothly.

A company relocation package can significantly ease the transition process. These packages often include a variety of benefits designed to help with the logistics and financial burden of moving. One such benefit is the provision of temporary housing.

Temporary housing allows you a place to live for a short period (typically 30-60 days) when you move to your new location. This can be extremely beneficial as it provides you with immediate accommodation while you get to know your new area, finalize the purchase or rental of your new home, or wait for your belongings to arrive.

This not only alleviates the pressure of finding a new home immediately but also provides stability as you transition into your new job and community. You have a place to rest, settle in, and focus on your new role while dealing with the other aspects of relocation. This buffer period can make the overall move less stressful and more organized.

However, it's crucial to clarify the terms of the temporary housing benefit with your new employer. Be sure to ask questions like: How long is the temporary housing period? What costs are covered? What happens if I can't find permanent housing within the provided timeframe? Understanding these details will help you plan accordingly and make the most of the benefits offered.

Final Career Preparation Checklist

o Job Offer in Hand: Make sure you have a written job offer before making final plans. This is crucial for various aspects, including securing a mortgage.

o Salary Negotiation: Be prepared to negotiate your salary. Remember to consider the total compensation package, not just the base salary.

o Understand Your Benefits: Know what benefits your new employer offers, including health insurance, retirement contributions, paid time off, etc. Be clear about when these benefits start.

o Relocation Package: Understand the terms of your relocation package if one is offered. Know what costs it covers and which ones you'll need to handle.

o Start Date Confirmed: Confirm the start date for your new job. Plan your relocation timeline around this.

o Contact Information: Make sure you have the contact information of your new HR representative or supervisor. Don't hesitate to reach out if you have any questions or concerns.

o   Company Culture: Do some research to understand the culture of your new workplace. This can help you adjust faster.

o   Resignation Letter: If you are currently employed in a civilian job, prepare your resignation letter, and know the proper procedure for submitting it.

o   Reference List: Prepare a list of professional references in case your new employer requests it.

o   Professional Attire: Depending on your new job, you might need to invest in a new wardrobe. Start shopping for professional attire suitable for your new role. Some companies may also provide a uniform or a clothing allowance for this purpose.

o   Continuous Learning: Consider any new skills you might need to excel in your new role and make a plan to acquire them.

Remember, it's a significant transition and it's okay to feel overwhelmed. Take one step at a time, reach out for help when you need it, and keep your eyes on the rewarding new career that awaits you.

6.4 The First Day of the Rest of Your Life: Starting Your New Job

Congratulations on making it this far in your civilian mission! It's been an incredible career transition road paved with hard work, dedication, and careful planning, and you've navigated it with resilience and determination. You've spent the last three years laying the groundwork, building your network, honing your skills, and setting clear career goals. Now, it's time to take those crucial steps into your new civilian career.

Starting your new job marks an exciting milestone in your journey. It's an opportunity to apply the knowledge and skills you've developed over the years and to contribute meaningfully in a new context. But remember, while this may feel like the end of your transition, it's truly just another beginning.

The transition from military to civilian life is not a one-time event. It's a process, one that will continue to unfold long after you've settled into your new career. This process of continual adaptation and learning is what will keep you agile and resilient in the face of future challenges and opportunities.

So, as you walk into your new role, remember to bring along the same tenacity, commitment, and adaptability that brought you this far.

Welcome to the first day of the rest of your life. Here's to making it count!

Here's a list of tips for making a good impression in a new job:

    1. Dress Appropriately: Dressing well shows that you respect your workplace and the people in it. Understand the dress code of your new workplace and follow it.

2. **Arrive Early**: Be on time or a bit early for your first day and establish this as a regular habit. Punctuality shows respect for others' time and demonstrates good time management.

3. **Positive Attitude**: A positive and enthusiastic attitude can make a strong impression. Show eagerness to learn and be open to new ideas.

4. **Active Listening**: Listen carefully during training and when receiving instructions. Ask clarifying questions if you don't understand something.

5. **Introduce Yourself**: Don't wait for others to introduce themselves to you. Be proactive in introducing yourself to your new colleagues.

6. **Ask Questions**: Don't hesitate to ask questions when you need clarity. It's better to ask than to guess and make unnecessary mistakes.

7. **Show Initiative**: Once you're comfortable in your role, show initiative by suggesting new ideas or volunteering for tasks. This shows your commitment and can make a positive impression.

8. **Stay Organized**: Keep your workspace tidy and manage your tasks efficiently. This can help you maintain productivity and also shows respect for shared workplace areas.

9. **Respectful Communication**: Treat everyone with kindness and respect, regardless of their role in the organization. Good manners never go unnoticed.

10. **Learn the Culture**: Every workplace has a unique culture. Take time to observe and understand the company culture, norms, and values.

Remember, first impressions are crucial, but they are just the start. Consistently demonstrating these behaviors can help you to continue making a positive impression as you move forward in your new job.

You'll be busy with many tasks as you start your new job, but it's also important to start a habit of managing your professional presence. A few housekeeping tips to accomplish in your first week:

1. **Update Your LinkedIn Profile**: With your new position confirmed, it's time to update your LinkedIn profile. Make sure your new job title, company, and start date are accurately represented. Share a post announcing your new role to let your network know about this exciting development.

2. **Pull Resumes from Job Boards**: As you step into your new role, remember to remove, or deactivate your active resumes from job boards and career sites. This will help prevent recruiters from reaching out about opportunities you're no longer available for, and signal that you've successfully transitioned into your new position.

3. **Update Your Social Media Bios**: If you have other professional social media accounts, such as Twitter or a professional Facebook page, make sure to update your bios there as well.

4. **Email Your Network**: Consider sending a brief, friendly email to your network announcing your new position and thanking them for their support during your job search. This is a great opportunity to express your gratitude and keep your contacts informed about your career progress.

5. **Start Strong in Your New Role**: Prepare yourself to make a great first impression. Do some additional research about your new company's

culture, revisit the job description to understand your responsibilities, and set some initial short-term goals for your first few months on the job.

6. **Revisit Your Personal Brand**: With this new role, consider how you want to evolve your personal brand. What new skills or experiences will you gain? How can you leverage your current brand to establish credibility in your new role?

Remember, the goal of these steps is not just to update your professional status, but to also lay the groundwork for your continued career growth.

Best of luck in your new role!

CHAPTER 7:
THE AFTERMATH - YOUR FIRST YEAR AS A CIVILIAN

After three years of careful preparation, you've finally stepped into your new career as a civilian. Congratulations on reaching this major milestone! The journey you embarked on with this guide has prepared you well for what lies ahead. Yet, as with any new journey, the road ahead may present both challenges and opportunities.

This first year in your civilian career will be instrumental in shaping your trajectory. According to a 2019 study conducted by the U.S. Bureau of Labor Statistics, veterans tend to have a lower unemployment rate compared to non-veterans. However, while this statistic is encouraging, it's important to remember that landing a job is just the first step.

Ensuring the longevity and success of your civilian career is crucial. A report from VetAdvisor and Syracuse University's Institute for Veterans and Military Families found that nearly half of veterans leave their first post-military job within a year, and approximately 65% do so within two years.

Why is this the case? The reasons vary. For some, their first civilian job might not be the right fit. Others may face challenges in adjusting to a different work culture. For others, career advancement opportunities might come their way.

But let's focus on you. This first year is an opportunity to apply everything you've learned from this guide, from networking to skills development, to personal branding, and more. Take this time to solidify your place in your new role and adjust to the civilian workplace culture. Understand that there will be differences and potential challenges, but remember, you've successfully navigated one of life's most significant transitions - from military service to civilian life.

Stay proactive in your career development. Keep your network alive and continue to learn and grow professionally. Seek feedback, and don't hesitate to ask for help when needed. Most importantly, remember to take care of your well-being. Balancing a new career with other life changes can be stressful, and it's important to prioritize your health.

As you navigate your first year as a civilian, remember that your military service has provided you with unique strengths and skills that are valuable in the civilian workforce. You've shown dedication, resilience, and adaptability in your military career. Now it's time to bring those same qualities to bear in your civilian career.

7.1 Adapting to Civilian Work Culture

Adapting to a civilian work culture is a significant aspect of your post-military career journey. As you transition into this new phase, one of the most vital elements to remember is that the civilian world operates differently from the military. The clear rank structure, uniforms signifying authority, and direct communication style that characterize military service may not always translate directly into a civilian work context. Understanding and respecting this difference is essential to your successful adaptation.

In the civilian work culture, hierarchy often exists, but it may be less visible and more fluid than in the military. You may find yourself in situations where the person with the most influence isn't necessarily the one with the highest rank or title.

Remember, you're no longer "wearing your rank" in the workplace. Titles like Master Sergeant, Lieutenant, Captain, or Colonel are not part of your professional identity in the civilian world. Instead, your skills, work ethic, and contributions to the team are what define you. It's essential to approach every interaction with humility and respect, irrespective of rank or status.

Keeping up with those who have also transitioned is important. Think again about the resources we provided in earlier chapters and use these groups to connect and share your experiences with others:
Some recommended sources include:

1. **LinkedIn's Veterans Mentor Network:** This is a community where military veterans and their spouses can ask career-related questions and share their experiences.

2. **Military.com's Transitioning Veteran group:** This forum has discussions about transitioning from military to civilian life.

Another significant shift you'll notice is in the realm of communication. Civilian workplaces often place a high value on collaboration, consensus-building, and indirect communication. This can be markedly different from the direct and command-oriented communication styles common in the military.

It's also important to keep in mind that progression in civilian jobs may not be as linear as in the military. There might be times when a lateral move - or even a step back - can open up more significant opportunities for growth and advancement in the future.

Transitioning into a civilian work culture might feel like learning a new language, but remember, you've already mastered one transition: the shift from civilian to military life. Now, you're just learning to switch codes when needed. The adaptability and resilience that saw you through your military service will serve you equally well in this context.

Finally, remember that while this transition can be challenging, it also represents an exciting opportunity for growth. With an open mind and a positive attitude, you can adapt to and even thrive in your new work culture. The skills, values, and experiences you bring as a military veteran are valuable, and they will help you make a meaningful contribution to any civilian workplace.

Here's a short list of strategies to help buffer the transition from military to civilian work culture:

1. **Keep an Open Mind:** Remember that the civilian world operates differently than the military. Be open to different ways of doing things and be patient with yourself and others as you adapt to your new environment.

2. **Communication is Key:** In many civilian workplaces, communication can be more indirect than in the military. Practice active listening and don't be afraid to ask questions if you're unsure about something.

3. **Build Relationships:** Develop professional relationships with your colleagues. This can help you understand the dynamics of your new workplace and also provide you with a support system as you navigate your transition.

4. **Seek Mentorship:** Find someone in your new workplace who can serve as a mentor. They can provide guidance as you navigate the norms and culture of your new job.

5. **Leverage Your Skills:** While the culture may be different, many of the skills you developed in the military, such as leadership, discipline, and problem-solving, are highly valued in the civilian world. Make sure to leverage these skills in your new role.

6. **Be Patient:** Transitioning to a new work culture takes time. Don't expect to understand and adapt to everything immediately. It's a learning process.

7. **Seek Support:** If you're finding the transition particularly challenging, seek support. This could be through a workplace support program, an external career coach, or groups for veterans transitioning to civilian life.

Remember, adapting to a new culture takes time, so be patient with yourself during this process. You're not alone in this journey, and some resources and people can help.

7.2 Making the Most of Your New Role: Performance and Promotion

Entering your new role with a proactive mindset can be key to succeeding in the civilian workforce.

According to Tom Peters, an American writer on business management practices, "Excellent firms don't believe in excellence - only in constant improvement and constant change." This implies that success doesn't come from merely striving for excellence but from continually improving and adapting to change - valuable advice for anyone transitioning from the military to the civilian workforce. In the same vein, Tony Robbins, a recognized authority on peak performance, suggests that "setting goals is the first step in turning the invisible into the visible." Set clear and achievable goals for yourself in your new role. Make them specific, measurable, achievable, relevant, and time-bound (SMART), to help you focus your efforts and make tangible progress. Furthermore, internationally acclaimed leadership authority, John C. Maxwell, believes that "leadership is not about titles, positions or flowcharts. It's about one life influencing another." In your new role, strive to positively influence those around you, regardless of your rank or position. This is a powerful way to make the most of your role and can open doors for promotion or advancement.

Your First-Year Performance Checklist

1.   **Set Clear Goals**: Write down specific, measurable, achievable, relevant, and time-bound (SMART – Chapter 3) goals for your role. This could include job-specific targets, learning objectives, or soft skills you want to develop.

2.   **Track Your Progress**: Keep a journal of your accomplishments, big and small. This will serve as a tangible record of your achievements and progress, which can be useful for performance reviews and future salary negotiations.

3. **Request Feedback**: Regularly ask for feedback from your supervisors and colleagues. This can provide valuable insights into your performance and highlight areas for improvement.

4. **Improve Continuously**: Identify areas where you can improve and take steps to develop in these areas. This could involve further training, attending workshops, or simply practicing a particular skill.

5. **Build Relationships**: Invest time in getting to know your colleagues and superiors. Attend social events, participate in team-building activities, and be available to help others when possible.

6. **Demonstrate Leadership**: Look for opportunities to demonstrate leadership, such as volunteering for projects or helping to solve problems. This can raise your profile within the company.

7. **Stay Healthy**: Look after your physical and mental health. Ensure you are getting enough sleep, eating healthily, exercising regularly, and taking time to relax and de-stress.

8. **Reflect**: Take time at the end of each month to reflect on your progress. Are you on track to achieve your goals? What challenges have you faced and how have you overcome them? What have you learned?

9. **Plan for the Future**: Towards the end of your first year, start thinking about your future with the company. What are your career aspirations? What steps will you need to take to achieve them?

Remember, transitioning to a civilian career is a process, and it's okay to take the time to adjust. Use this checklist as a guide to help you navigate your first year and beyond.

What about Promotions?

Promotions in the civilian corporate world can look quite different from those in the military or federal jobs. In the military and federal jobs, promotions often follow a structured timeline and are based largely on time in service and successful completion of training and performance evaluations. The path to promotion is clear and somewhat predictable.

In the civilian corporate world, promotions can be less regimented and more dependent on a variety of factors. While performance evaluations, skills, and experience do play a significant role, promotions may also be influenced by the company's needs, budget, and internal politics. A position may not open up until someone leaves, or a new role may be created to meet changing business needs.

Here are some key points to keep in mind:

Metrics and Your Role: In the civilian corporate sector, your performance plays a significant role in determining your promotion potential. It's important to start viewing your contributions not just as tasks or projects completed, but as measurable impacts on the business. This could be in terms of increased sales, cost savings, time efficiencies, or improvements in key performance indicators (KPIs). Using numbers, dollar signs, and percentages to quantify your achievements can provide a clearer picture of your value to the company.

For instance, rather than saying you "led a team project," you could say you "spearheaded a team of 5 to complete a project 3 weeks ahead of schedule, saving an estimated $20,000 in project-related costs." This concrete data can strengthen your case when discussing promotions or salary increases with your superiors. Remember, your goal is to demonstrate the tangible value you bring to the company, and hard numbers speak volumes.

Building Relationships is Crucial: Cultivating Mentors, Sponsors, and Champions

At the beginning of this book, in the Acknowledgements section, I thanked my mentors, sponsors and advocates (champions). I gained relationships with each one of these people through my professional work as a consultant, an employee, or a civic volunteer.

Networking and building strong relationships within your organization and your industry are essential steps for career advancement. This should include not only peers and superiors, but also mentors, sponsors, and champions.

Mentors are individuals who can guide you based on their own experiences and knowledge, helping you navigate your career path with the wisdom and insights they've gained over the years. They often work in a similar field or role as you and can provide valuable advice and perspective.

Sponsors are individuals who have influence within your organization or industry and are willing to use that influence to advocate on your behalf. They can help you gain visibility and provide opportunities for advancement that you might not have access to otherwise.

Champions are individuals who believe in your potential and will speak highly of you to others. They might be colleagues, clients, or even managers who you've impressed with your skills and work ethic.

All three roles are valuable, and you may find that one person can fulfill multiple roles. It's important to nurture these relationships over time, as they can provide support, offer opportunities, and help you navigate the challenges of your new civilian career.

Skill Development is Key: Pursuing Continuous Learning and Growth: Continually improving your skills and staying updated with industry trends can make you a more valuable asset to the company, enhancing your promotion potential. Taking on challenging projects or assignments that are outside your comfort zone can also help showcase your capabilities.

153

In the civilian world, just as in the military, lifelong learning is a key component of career advancement. Many companies recognize this and offer tuition assistance or education reimbursement programs to encourage their employees to further their education and acquire new skills. These programs might cover all or part of the cost of degree programs, certification courses, or professional development seminars.

Do note that while many companies offer this benefit, there may be stipulations such as staying with the company for a certain period after completing the program.

As a veteran, you also have access to education benefits like the post-9/11 GI Bill, which can cover degree programs, vocational, and technical training. These resources can provide you with additional avenues for improving your skills and knowledge, making you an even more valuable asset in your new civilian career. In sum, continuous learning not only increases your promotion potential but also keeps you adaptable in a constantly changing work environment.

Promotions May Not Follow a Set Timeline

Unlike the military, there may not be a set timeline for promotions in the civilian sector. It can happen within a few months, or it may take several years. Much depends on the individual company's structure, needs, and culture.

To navigate this unfamiliar territory, it's important to maintain open lines of communication with your manager and others who can support your growth. Regular check-ins about your performance and career goals can provide useful feedback and align expectations.

If your company doesn't have a formal system for setting and reviewing goals, consider taking the initiative to establish your own short-term and long-term career objectives. Discuss these with your manager, seeking their input and support. This proactive approach demonstrates your commitment to your role and your future with the company.

By fostering these relationships and being proactive in your career development, you can better position yourself for advancement opportunities when they arise, even if they don't follow a specific timeline.

Negotiating is Often Necessary

Unlike in the military, civilian career growth often involves advocating for yourself and negotiating for promotions or salary increases. This can seem intimidating, especially if you're unfamiliar with the process, but remember the concepts of anchors and perceived value that we discussed in previous chapters.

Your anchors are your non-negotiables, such as a specific salary range based on your research of market rates for your role and location, or the need for a certain level of responsibility and challenge in your work. Your perceived value is the unique combination of skills, experience, and potential you bring to your role, which you have been developing and demonstrating throughout your time in the company.

Here are some tips for successful negotiation in a current job:

1. Document Your Achievements: Regularly update a list of your accomplishments in your role, quantifying the impact of your work whenever possible. Use this as evidence of your value to the company during negotiation discussions.

2. Understand Market Rates: Keep up to date with the salary range for your role in your industry and location. Resources like Glassdoor, Payscale, and the Bureau of Labor Statistics can provide useful data. If you're seeking a raise, it's crucial to know what a competitive salary for your role and experience level looks like.

3. Prepare Your Case: Before approaching a negotiation, prepare a clear case outlining why you deserve a promotion or raise. Use your list of achievements and market data to support your argument.

4. Practice Your Pitch: Negotiation is a skill that improves with practice. Role-play the conversation with a trusted friend or mentor and seek feedback on your pitch.

5. Be Prepared to Listen: During the negotiation, listen to your employer's perspective and be prepared to respond to any concerns or counteroffers they might present. Remember that negotiation is a two-way conversation, and both parties should walk away feeling satisfied with the outcome.

6. Review and Understand Your Benefits: While salary is a crucial part of your compensation, also consider benefits like health insurance, retirement contributions, and professional development opportunities. A cost-of-living, or market salary adjustment or a title change might also be part of the negotiation.

Remember, negotiation is a normal part of professional growth. It's a conversation, not a confrontation, and it shows your commitment to your role and your career.

Understanding these differences can help you better navigate the civilian corporate promotion process. It can require patience, strategic planning, and proactive career management. Remember that career advancement isn't always linear in the corporate world, and sometimes lateral moves can offer opportunities for skill growth and future promotions.

7.3 Evaluating Your Fit: When to Stay and When to Leave Your New Role

How can you evaluate if your current job is serving you well? It's not uncommon for veterans in transition to initially choose a job or career path that ultimately doesn't align with their values, interests, or long-term goals. You may feel a strong commitment to stick it out, especially if it's your first civilian role, but it's essential to consider your happiness, career aspirations, and mental health.

Here are some clear signs that your current job may not be a good fit:

1.	Feeling Consistently Stressed or Unhappy: Work can be challenging and stressful at times, but if you're regularly feeling overwhelmed, anxious, or unhappy, it may be a sign that the job isn't right for you.

2.	Misalignment with Values: If the company culture or the nature of your work doesn't align with your values, it can lead to dissatisfaction and a sense of being out of place.

3.	Lack of Growth Opportunities: If you're not seeing opportunities for personal growth or career progression, or if your job doesn't offer you avenues to learn new skills and expand your knowledge, it may not be the right fit for your career goals.

4.	Toxic Workplace Environment: Signs of a toxic workplace may include a lack of respect or support from management, unethical practices, consistent negativity, or an overly competitive environment that stifles teamwork and collaboration.

5.	Physical or Mental Health Concerns: If your job is negatively affecting your physical or mental health, it's crucial to take action. This can be through seeking support from health professionals, speaking to HR or management about your concerns, or considering whether a different job may be a healthier choice.

6.   Utilization of Prior Military Background for Dangerous Tasks: Unfortunately, some employers may take advantage of the perceived "agentic" qualities and former training of military veterans, assigning them to tasks or roles that are potentially dangerous without proper training or risk mitigation strategies. If you find yourself in a situation where you believe your employer is unfairly or unsafely capitalizing on your military background, it's essential to speak up. Communicate your concerns to your supervisor or HR, and if necessary, consider seeking advice from a legal professional or a veterans' advocacy group. Remember, your safety and well-being are paramount, and no job should put you at undue risk. Ever.

Important to note the word "agentic" is a term used to describe individuals who are self-assertive, self-confident, persistent, and likely to take charge in situations. These are, of course, valuable traits, particularly in a military setting where clear, decisive action is often necessary.

However, in the civilian work culture, which often values collaboration, consensus-building, and more nuanced approaches to problem-solving and leadership, these traits can be misunderstood or misapplied.

Some common perceptions that civilians may have about military members include being overly hierarchical, rigid, or autocratic in their leadership style. They may be viewed as always preferring a direct, even blunt, communication style. Civilians may also have a perception that veterans aren't flexible or adaptive, as they're coming from a highly structured environment with clearly defined roles, rules, and procedures.

The potential mismatch between these military-acquired traits and civilian work culture expectations can lead to misunderstandings and miscommunications. For example, a veteran's direct communication style may be perceived as brusque or insensitive in a civilian workplace that values softer, more diplomatic communication. Similarly, a preference for clear hierarchies and definitive decision-making can seem out of place in environments that value team consensus or a more democratic style of leadership.

The issue here is not with a veteran's skills or capabilities; rather, it's a matter of understanding the different norms and expectations in civilian workplaces and learning to navigate them effectively. Military veterans have a wealth of experience and a host of valuable skills that can greatly benefit civilian organizations, but both veterans and their potential employers need to understand and bridge the culture gap for a successful transition.

Helping employers and colleagues understand the value that veterans bring to the table, while also helping veterans adapt their skills and experiences to the civilian work culture, is critical. This is why transition assistance, mentorship, and ongoing support are so important in facilitating a successful military-to-civilian career transition.

Need A Do-Over? Go Right Ahead.

As you've probably gathered by now, the journey of transition from military to civilian career as outlined in this book is cyclical. If the first job after military service isn't a good fit, it's entirely acceptable and often beneficial to revisit certain chapters of this book to guide the next steps.

Here's how to use this book for the next round of job search:

1. **Reflect on the Experience:** Revisit Chapter 7 and consider what didn't work in your previous role. Was it the culture? The type of work? The industry? This reflection will provide invaluable insights for your next job search.

2.   Update Your Career Goals: Based on your recent experience, you may need to refine your career goals. Review Chapter 1 and update your goals as needed.

3.   Improve Your Resume and LinkedIn Profile: Use the advice in Chapter 3 to revise your resume and LinkedIn profile to reflect your latest job experience and the new skills you have gained.

4.   Networking: Continue to build your civilian network. Remember the strategies in Chapter 2 on growing your network both online and offline.

5.   Job Search: Once your goals are updated and your materials ready, return to Chapter 4 for advice on applying for jobs and target companies in your desired industry.

6.   Salary Negotiation: Revisit the salary negotiation strategies in Chapter 4 and apply these to any job offers you receive.

7.   Start the New Job: When you land a new role, go back to Chapter 6 for guidance on starting the new job effectively.

Remember, career transition is not a linear process. It's a journey, and each experience, even if not perfect, brings you one step closer to finding the right fit. Don't be discouraged if the first attempt doesn't work out as planned. It's all part of the process. Each experience is an opportunity to learn more about your strengths, preferences, and the kind of work that brings you fulfillment.

In essence, the transition is a process, not an event. Even years after putting away their uniform, veterans say they're still trying to navigate their changed lives and make the shift from being service members to civilians.

CHAPTER 8:
LONG-TERM SUCCESS - 5+ YEARS AFTER THE MILITARY

As you cross the five-year mark in your civilian career, reflect on where you are, and where you came from. The five-year mark is a good bearing on where your civilian career can take you. You are no longer in 'survival mode" but on a steady course. The transition from military to civilian life is a journey, and long-term success is about maintaining the momentum you've built. It's time to reflect on the past, evaluate the present, and strategize for the future.

Reflecting on the Journey

The first step to understanding your long-term success is to look back at where you began. You traded your uniform for business attire, your military jargon for corporate communication, and your structured chain of command for a more nuanced matrix of relationships. The skills and disciplines that were honed in the service have now become integral parts of your civilian professional persona.

Continued Professional Development

Education doesn't stop. The landscape of business and technology is constantly evolving, and so should you. Continued professional development may mean pursuing further education, additional certifications, or leadership training. Stay ahead of industry trends and be proactive about learning new skills that will make you an invaluable asset to any team or project.

Financial Planning and Growth

Long-term success isn't just about job titles; it's also about financial stability and growth. Hopefully, you've been able to leverage your skills into a position that not only satisfies you professionally but also rewards you financially. Now is the time to review your salary and ensure that it is in line with your goal setting and expectations.

Networking and Relationships

The network you began building in the early stages of your transition should continue to evolve. Continue to nurture the relationships you've built and expand your network. Remember, networking is a two-way street. Continue to offer support to those in your network, and don't hesitate to reach out when you need guidance or opportunities.

Giving Back

One of the most rewarding aspects of long-term success is the ability to give back. This could mean helping those who are now standing where you once were—transitioning veterans. Consider volunteering, mentoring, or engaging in speaking opportunities to share your story and insights.

Preparing for the Next Transition

Just as you once transitioned out of the military, there will come a time when you'll face another transition, whether it's retiring from your civilian career, starting a business, or changing your professional course. Start preparing for this next phase with the same strategic planning that served you in your military-to-civilian transition.

The Long View

Long-term success is a continuous process of personal and professional development. It is about being adaptable, just as you were in the military. You have proven your resilience and capacity to grow, and the next five years—and beyond—are canvases for you to create even more success.

Remember, the discipline, adaptability, and leadership you developed in the military are the foundations of your civilian achievements. As you continue to build upon them, there's no limit to the success you can attain.

Exercise: Milestone Review
Take a moment to write down the significant milestones since your transition.

- What skills have you acquired?
- Which goals have you achieved?
- How have your military experiences influenced your path?

This reflection is crucial—it's your personal history of success that will propel you forward.

Evaluate the Present
Now, assess where you stand today.

- You've likely climbed the ranks, perhaps transitioning from a team member to a leader.
- You've expanded your skill set, network, and maybe even changed industries or roles.

Evaluating the present means understanding your current position, job satisfaction, work-life balance, and personal growth.

Ask yourself these questions:

- How closely does my current career align with my initial goals post-transition?
- What have I learned about my professional desires and abilities in the last five years?
- Have I continued to grow and challenge myself?
- Am I mentoring or assisting others in their transition journey?

As you close this chapter, take a moment to appreciate how far you've come and get excited about where you're going. Your Civilian Mission continues, and this career journey is far from over—it's just entering a new phase.

8.1 Stories of Success: Veterans in Businesses, Corporations, Leadership, and Entrepreneurs

Military transition is not a one-time event, it's a lifelong journey. The skills, values, and experiences you've acquired during service stay with you, shaping your perspective and approach to civilian life. This journey is not about leaving the military behind, but about learning to integrate that part of yourselves into your new roles and identities.

Many celebrities, entrepreneurs, and business leaders have served in the military. Here are a few notable examples:

1.   Ice-T (Musician, Actor): Ice-T, whose real name is Tracy Marrow, served in the U.S. Army for four years in the 25th Infantry Division, before starting his music and acting career.

2.   Bea Arthur (Actor): The "Golden Girls" star was one of the first members of the U.S. Marine Corps Women's Reserve who rose to the rank of Staff Sargeant having served during World War II.

3.   Rob Riggle (Comedian, Actor): Known for his roles in "The Hangover" and "Step Brothers," Rob Riggle served in the U.S. Marine Corps Reserve, retiring as a Lieutenant Colonel.

4.   Frederick W. Smith (Founder of FedEx): Before he founded FedEx, Frederick W. Smith served in the U.S. Marine Corps as a logistician, serving two tours of duty during the Vietnam War.

5.   Paul Bucha (Businessman): Paul Bucha, a Vietnam War Veteran and Medal of Honor recipient from the U.S. Army, has held leadership roles at multiple corporations and serves on the boards of several others.

6.   Montel Williams (TV Personality): Before his TV career, Montel Williams enlisted in the Marine Corps and later, became the first Black man to graduate from the Naval Academy.

7.   Roger Staubach (Former NFL player, Real Estate Entrepreneur): After graduating from the Naval Academy

where he won the 1963 Heisman Trophy, and serving in the Navy, Roger Staubach went on to a successful athletic career in the NFL before becoming a successful entrepreneur in the real estate industry.

8. Harvey C. Barnum Jr. (Businessman): Harvey C. Barnum Jr., a Medal of Honor recipient, and U.S. Marine Corps veteran, went on to hold several corporate leadership roles after his military service.

9. Pat Sajak (TV Personality): The "Wheel of Fortune" host served in the U.S. Army as a disc jockey during the Vietnam War.

10. Wes Moore (Author, CEO): Moore is a U.S. Army veteran who served as a paratrooper and Captain with the 82nd Airborne in Afghanistan. After his military service, Moore authored "The Other Wes Moore," a New York Times bestseller, and became the CEO of Robin Hood, one of the largest anti-poverty nonprofits in the United States.

11. J.R. Martinez (Actor, Motivational Speaker): Martinez was severely injured while serving in the U.S. Army in Iraq, suffering burns over 34% of his body. After his recovery, he became an actor, appearing on "All My Children," and won the 13th season of "Dancing with the Stars."

12. Shilo Harris (Author, Motivational Speaker): Harris was severely burned by a roadside bomb during his second deployment in Iraq. After a long recovery, Harris became a motivational speaker and published a memoir titled "Steel Will: My Journey through Hell to Become the Man I Was Meant to Be."

13. Paul Rieckhoff (Author, Podcast Host, and Founder of Iraq and Afghanistan Veterans of America): Rieckhoff served as a First Lieutenant and infantry rifle platoon leader in the U.S. Army in Iraq. After his service, he founded the Iraq and Afghanistan Veterans of America (IAVA) and hosts the "Angry Americans" podcast.

14. Emily Miller (Corporate Executive): After serving in the U.S. Army, Miller worked for General Electric in various

leadership roles and is now the Vice President of Operations at Arrow Electronics, a major technology company.

15. Adam Driver (Actor and Producer): After serving the US Marine Corps, Driver led a group of actors through a series of plays and performances at various USMC installations in the US. He has been nominated for two Academy Awards, four Primetime Emmy Awards, and a Tony Award.

These examples are a testament to the fact that veterans can and do succeed in a wide array of fields after their military service. The skills, discipline, leadership, and resilience developed in the military often translate well to civilian life and can set veterans apart in their post-military careers.

Here are successful veterans who have made significant contributions in their field(s) post-military:

1. "Leadership in Action" by Richard Marcinko: Marcinko, the founder and first commanding officer of SEAL Team Six, takes readers into the world of the Navy's elite force and explains how the lessons he learned in the SEALs can be applied to the business world.

2. "Make Your Bed: Little Things That Can Change Your Life...And Maybe the World" by Admiral William H. McRaven: In his book, McRaven shares the ten principles he learned during Navy SEAL training that helped him overcome challenges not only in his long Naval career but also throughout his life.

3. "Extreme Ownership: How U.S. Navy SEALs Lead and Win" by Jocko Willink and Leif Babin: Willink and Babin apply lessons learned in combat, explaining how the principles that enable SEAL units to achieve their best on the battlefield can help us achieve success in our personal and professional lives.

4.   "Call Sign Chaos: Learning to Lead" by Jim Mattis: This book by the former Defense Secretary and retired Marine Corps General Jim Mattis recounts his career and imparts leadership advice learned from his decades in the military.

5.   "Beyond the Call: Three Women on the Front Lines in Afghanistan" by Eileen Rivers: Eileen Rivers, a USA Today editor and former Army journalist, uses her personal experiences and those of three women in the military to showcase the valiant contributions women have made in recent military conflicts.

6.   "Hesitation Kills: A Female Marine Officer's Combat Experience in Iraq" by Jane Blair: Jane Blair, a former U.S. Marine Corps Reserve officer, discusses her combat experiences in Iraq and the challenges and opportunities of being a female service member in a war zone.

7.   "Find Another Dream" by Maysoon Zayid: Maysoon Zayid, a former U.S. Air Force Reserve officer, shares her journey from growing up as a disabled Palestinian Muslim in New Jersey to becoming one of America's first Muslim women comedians.

8.   "The Ellipsis Manual" by Chase Hughes (who also wrote the foreword to this book), a former Navy officer who took 20 years of HUMINT training to write a best-selling book, just three years after his military service ended. He continues to write books and stay on the best-seller lists while training LEO and Government Agencies, as well as civilians, on body language, behavior, and neuroscience.

9.   "The Body Language Handbook" and 11 other titles come from the extensive writings of former U.S. Army interrogator and SERE trainer, Greg Hartley. Greg is the co-founder of The Behavior Panel on YouTube (along with fellow veteran Chase Hughes) which has gained hundreds of thousands of weekly viewers.

These veterans have found success in their post-military careers and their books offer valuable insights on the lessons learned from military service. However, the applicability of military experiences varies among individuals and career paths. It's important to note that these examples may not directly quote the authors' successful post-military career transitions, but they illustrate how military training and experience can contribute to success in a variety of fields.

Vet-repreneur?

There's no denying that transitioning from military to civilian life and navigating the waters of corporate, federal, or civilian employment can be challenging. It requires adapting to new cultures, learning new ways of operating, and in many cases, reshaping the skills you've honed during your military career into a form that's recognizable in a civilian context.

But what if these traditional paths of employment don't feel like the right fit for you? What if you have a burning desire to be your boss or a brilliant business idea that you're eager to bring to life? What if you want to take control of your career path and carve out a road that's uniquely yours?

That's where entrepreneurship comes in. In the world of entrepreneurship, the discipline, leadership skills, and problem-solving abilities you gained in the military can serve as powerful assets. The potential to create something meaningful, generate jobs, and contribute to the economy can be incredibly rewarding. Plus, as an entrepreneur, you have the opportunity to build a work culture that respects and values the skills and experiences veterans bring to the table.

Transitioning to civilian life doesn't have to mean fitting into a pre-existing mold - it can also mean creating a mold that fits you. Read on to learn about fellow veterans who've taken this path and turned their entrepreneurial dreams into reality.

Here's a list of successful veteran entrepreneurs, also known as "vetrepreneurs", who have built notable nonprofit businesses after their military service:

Non-Profit Veteran Entrepreneurs:

1. Mike Erwin (Former Army Officer) - Founded Team Red, White & Blue, a nonprofit connecting veterans to their community.

2. Sandra Edwards (Former Army Officer) and Beth Kluttz (Former Army Officer) - Co-founded Dog Tag Furniture, a nonprofit supporting family of fallen military personnel.

3. Emily McMahan (Veteran) and Sam Pressler (Veteran) - Run Capitol Post, a nonprofit teaching entrepreneurship to veterans and spouses.

4. Fred Wellman (Retired Army Officer) - Founded ScoutComms, a B-Corporation focused on veteran and military issues.

5. Steve Brown (Former Navy SEAL) - Founded Camp Brown Bear, a refuge and retreat for veterans with TBI, MSA, and PTSD.

6. Dan Rooney (F-16 Fighter Pilot in the Oklahoma Air National Guard) - Founded Folds of Honor, a nonprofit providing scholarships to children and spouses of fallen and disabled service members.

These veteran entrepreneurs have taken their military experiences and translated them into successful business ventures, serving as an inspiration to others following a similar path. Some have amassed millions of dollars of revenue.

For-Profit Veteran Entrepreneurs:

1. Nick Taranto (Former Marine Corps) - Co-founded Plated, a meal kit service (later sold to Albertsons).

2. Phyllis Newhouse (Retired Army Officer) - CEO of Xtreme Solutions (IT services and cybersecurity) and co-founder of ShoulderUp (women entrepreneurs' platform).

3.   Brandon Shelton (Former Army Officer) - Founder of Task Force X Capital, a venture capital firm investing in veteran-led businesses.

4.   Chase Hughes (Former Navy HUMINT Officer) - Founder of Applied Behavior Research (coaching and training company) and co-host of The Behavior Panel on YouTube.

5.   William Allen (U.S. Marine) - Founder and Managing Partner of 39Alpha (early-stage venture capital firm).

6.   Kevin Seiff (Former Navy SEAL) - Founder of Flintlock Solutions (early and growth-phase start-up partner in tech and life sciences) and Founder of The Vetrepreneur Collective (training veterans for entrepreneurship).

7.   Robert D. Smith (U.S. Air Force Veteran) - Founder, chairperson, and CEO of Vista Equity Partners (private equity firm specializing in technology and software sectors).

8.   Fred Smith (Former Marine Corps) - Founder of FedEx, a multinational delivery services company.

9.   Andrew G. Cherng (U.S. Army Veteran) - Co-founder of Panda Express, a popular fast-food chain serving American Chinese cuisine.

These veterans successfully transferred the skills and experiences they gained in the military into the business world, illustrating the possibilities for post-transition success with determination and entrepreneurial spirit.

Remember, while these individuals have found success in their respective fields, each veteran's path is unique. Your journey may look different, and that's perfectly okay. The skills and experiences you gained in the military are valuable and can be translated into a wide variety of civilian careers.

8.2 Career Growth and Long-Term Planning

As you settle into your new role and the immediate challenges of transition subside, you'll want to start looking ahead. In this chapter, we'll talk about the importance of ongoing professional development and career planning. We'll discuss setting long-term career goals, seeking leadership and learning opportunities, and navigating career progression in the civilian sector.

Here's a targeted checklist based on the themes we've discussed throughout the book:

1. **Identify your long-term career goals:** Where do you want to be in five years? In ten years? What types of roles are you interested in? What skills will you need to develop to reach these goals?

2. **Continue to build your brand:** Maintain an active presence on professional networking platforms like LinkedIn. Regularly share insights and engage with others in your industry.

3. **Pursue continuous learning and professional development:** Seek out relevant courses, certifications, or degrees that can enhance your skills and knowledge.

4. **Stay updated on industry trends:** Follow industry news, attend conferences, and webinars, and keep learning about your field.

5. **Seek leadership opportunities:** Whether it's taking the lead on a project at work, volunteering for a leadership role in a professional organization, or starting your own business, these experiences can be valuable for your career growth.

6. **Maintain and expand your professional network:** Networking isn't just for job hunting. Regularly engage with

your network and look for opportunities to meet new people in your industry.

7. Periodically review and update your resume: Even if you're not job hunting, it's good practice to keep your resume updated with your latest accomplishments.

8. Ask for feedback and take constructive criticism: Regularly ask for feedback from your superiors, peers, and subordinates. This can help you identify areas for improvement and growth.

9. Don't forget to negotiate: When opportunities for promotion or salary increases arise, remember the value of negotiation.

10. Practice work-life balance: Prioritize your mental and physical health and maintain a healthy balance between your work and personal life.

Remember, career growth and planning are ongoing processes, and it's okay to adjust your goals and plans as your circumstances change and you grow as a professional.

8.3 The Importance of Giving Back: Mentoring Others in Transition

You've walked the path from military service to civilian career. Now, consider using your experience to help others navigate their transitions. In this chapter, we'll explore the value and satisfaction that can come from mentoring others and provide tips on how to be an effective mentor.

Mentoring others in transition is a powerful way of giving back and continuing the legacy of service beyond the military. Mentors who have navigated the transition from military to civilian life possess invaluable firsthand experience and insights that can greatly assist those who are embarking on the same journey. Moreover, veterans often share a unique bond and understanding, making them especially effective mentors for their fellow veterans.

According to the U.S. Department of Veterans Affairs, veteran mentors can greatly contribute to the successful reintegration of other veterans into civilian life. A study funded by the VA found that veterans who received peer support were more likely to use VA health services, less likely to be hospitalized, and more likely to show improvements in mental health symptoms and social functioning[1].

In addition, numerous testimonies speak to the positive impact of veteran mentorship. For instance, Kevin O'Brien, a former U.S. Marine Corps officer and now a successful entrepreneur, shared his views on mentorship in a LinkedIn post: "Mentorship has been a cornerstone of my life, especially during the transition from the Marine Corps. The importance of having a mentor who 'has been there' cannot be overstated."

Another veteran, Sean Kelley, who transitioned from the U.S. Navy to become a tech industry leader, emphasizes the importance of mentorship in an article for Task & Purpose: "As a veteran, one of the best things you can do is find a mentor who can guide you through the transition... Mentors can give you insight into opportunities you may have overlooked."

Other noteworthy reminders:

1. "The most effective tool you have in your transition arsenal is the ability to adapt—use it liberally. The second most effective tool is to teach others, use it wisely." – Retired O6 Colonel, R. M. Smith, Mentor at ACP

2. "Translate your military discipline into civilian drive, and you'll find the path to success is more similar than you think. When there, remind yourself of the journey, and give back to those looking for the light at the end of the tunnel." – Former E8 Master Sergeant, J. T. Lee, Transition Coach

3. "Your service may end, but the mission continues. Find your next purpose with the same conviction you served as a volunteer or mentor for other service members." – Retired O5 Commander, A. B. Nguyen, Veteran Career Advisor

4. "Leverage the leadership forged in your military career to elevate your civilian role. It's your most transferable skill, right next to your ability to give it openly and freely to those who need it." – Ex-O3 Captain, S. K. Patel, Professional Speaker

5. "In the military, we adapt and overcome. In civilian life, we apply and excel. Carry that ethos into your next chapter and shine the light so others can find the path." – Former E5 Sergeant, D. E. Garcia, Corporate Mentor

6. "Networking in civilian life is just like building your squad in the service. It's all about trust, support, and mutual goals. Find those who are following you and extend support. Be the person you wish you had before your military transition." – Retired O4 Major, L. C. Johnson, Networking Expert

7. "Remember that your military experience has equipped you with a global perspective—a valuable

asset in today's interconnected world. Don't limit yourself to set beliefs or only a specific group of like-minded individuals. Give back and grow in diverse ways." – Retired O3 Lieutenant, K. H. Wright, Career Strategist

8. "The camaraderie of the armed forces can be found in the civilian workforce; you just have to build it. Start with your new team. Be the builder." – Former E6 Staff Sergeant, G. F. Ellis, HR Consultant

9. "Embrace the journey from uniform to business attire as a natural progression of your professional life. Consider mentoring as your next professional hobby. By giving back, your potential is limitless." – Retired O5 Lt. Colonel, P. Q. Thomson, Business Development Manager

10. "Mentorship doesn't end with the service; it becomes part of your legacy. Continue to guide, lead, and inspire." – Former E7 Chief Petty Officer, N. R. Gomez, Veteran Outreach Coordinator

These quotes highlight the importance and the power of mentorship in helping veterans successfully navigate their post-military careers. By giving back through mentorship, veterans can continue to serve their community, contribute to the success of their fellow veterans, and reinforce the strong network of support within the veteran community.

CHAPTER 9:
CONCLUSION

9.1 The Lifelong Journey: Continuous Growth and Development

The transition from military to civilian life is not a single event, but a continuous journey of personal and professional development. As you embark on this journey, it's crucial to understand that it extends far beyond the initial transition period.

Like any significant life change, it's an ongoing process, punctuated by moments of introspection, learning, growth, and, yes, sometimes even setbacks. But it's these experiences that shape your journey, mold your character, and ultimately, contribute to your success in the civilian world.

Let's think of it as a voyage. In the military, you've been equipped with a unique set of skills, values, and experiences. These constitute your navigational tools and your vessel. Now, as you set off into the vast ocean of civilian life, you're embarking on an adventure filled with opportunities, surprises, and challenges. Your military training serves as your compass, but you'll need to chart your course and adjust your sails to the changing winds of civilian life.

Military service imparts numerous transferable skills, such as leadership, teamwork, problem-solving, and resilience. These skills, coupled with a disciplined mindset, are highly valuable assets in the civilian sector. But transitioning to civilian life often entails more than just leveraging these skills. It involves learning new norms, adapting to different work cultures, and sometimes even redefining your identity outside the uniform.

Experts caution that redefining your identity to fit a corporate culture has risks. Carin Richelle Sendra is a Post 9/11 USAF Veteran that served both in active duty as well as

D.O.D security contracting OCONUS. She has spent time in both the private and public sectors and shared in her article, "Veterans in the Civilian Job Sector: Stand Out or Blend In?" She writes, "From my experience, trying to "fit in" to a culture that doesn't feel correct for me, never works out very well in the end. I find myself suffering in silence as I contemplate having a steady paycheck versus valuing my voice and authenticity. I believe that transitioning service members may find themselves standing at the crossroads of having to choose between a paycheck or self-fulfillment early on in their career transition."

Having a growth mindset is particularly relevant as you navigate your career in the civilian world. Whether it's learning new skills, adapting to different work cultures, or taking on unfamiliar roles, each challenge offers an opportunity for growth and development.

Dr. Carol Dweck, a renowned psychologist, and author of the book "Mindset: The New Psychology of Success," reinforces the importance of the 'growth mindset' vs. a fixed mindset. She writes, " "In the fixed mindset, everything is about the outcome. If you fail—or if you're not the best—it's all been wasted. The growth mindset allows people to value what they're doing regardless of the outcome. They're tackling problems, charting new courses, working on important issues."

But as you journey through this transition, it's also essential to acknowledge the emotional aspect of this process. It's a significant life change, and it's okay to experience a range of emotions. It's okay to feel overwhelmed at times, and it's okay to ask for help.

One of the ways you can do this is to ensure that you are feeling comfortable in your work environment, Daniel Zia Joseph, author of "Backpack to Rucksack: Insight Into Leadership and Resilience From Military Experts" discussed "Command Climate" on The Art of Manliness Podcast; Leadership Lessons from Military Mentors. "Essentially, it's work culture. In the military, it's unique because the person at the top sets the temp for the entire unit. If you have someone who is super fit, all of the subordinates are fit. Conversely, if you have someone who is negative, or toxic, you'll see people below them treat others the same way. In the Military you can't change the "Command Climate." What you get is what you get. I've seen good and bad in the Corp environment. What is astounding to me is the way that it compounds the stress around us. Good leaders are enablers of subordinates to find solutions and figure out a way to make things better. "

Your transition journey does not end when you land your first civilian job or even when you make your first career move in the civilian sector. It's a lifelong journey that unfolds with each step you take, each decision you make, and each challenge you overcome.

Embrace the journey. Keep learning, keep growing, and remember that you're not alone.

You are part of a community of veterans who have walked this path before, and those who are walking it alongside you.

Lean on this community, learn from it, contribute to it, and together, we'll navigate the transition, achieve our career goals, and continue to serve in new and meaningful ways.

9.2 Final Words of Encouragement

As we reach the end of this guide, I want to extend my deepest gratitude to you. Your dedication, your service, and your commitment to our country and our freedoms are immeasurable. It's been a privilege and an honor to accompany you on this journey, from your days in uniform to your new life in the civilian sector.

I commend you for taking the bold step to chart a new course for your life and your career. And I thank you for allowing me, through this guide, to be a part of that journey. Your journey is a testament to your resilience, your adaptability, and your unwavering commitment to growth and development.

I've always believed that our lives are defined not by what we take from the world, but by what we give back. And you, dear reader, have given so much. You've dedicated years of your life to serving our country, and now you're forging a new path in the civilian sector, bringing your unique skills, experiences, and perspectives to enrich our workplaces and our communities.

As you continue on this journey, remember this: the learning never stops, the growth never ends, and the transition, in many ways, is a lifelong process. But with every step you take, with every challenge you overcome, you're adding a new chapter to your unique story.

I hope that this guide has served as a valuable resource for you. But more than that, I hope it has sparked a fire in you to continue learning, growing, and contributing to those around you.

In the words of Les Brown, motivational speaker, and author of "The Greatness Within You: Believe in Yourself and Discover Your Potential," He writes," You have more potential than you can ever begin to imagine. Most of us go to our graves 'holding on' rather than releasing our potential. Don't underestimate yourself. Don't downplay your abilities."

As you navigate your career and life beyond the military, I hope you'll continue to leverage your strengths, capitalize on your experiences, and make your mark in the world in your unique way. And as you move forward, I encourage you to reach back and help others on their transition journey. Share your experiences, lend your insights, and be the guiding light for others following in your footsteps.

As we part ways, I want to leave you with a few words of encouragement:

Continue to strive, to learn, to grow.
Believe in yourself and your abilities.
Embrace the journey, with all its twists and turns.
Ask for help when you need it.
And know that your military service is not just a part of your past,
but the building blocks and cornerstone for your future, a testament
to your strength, your resilience, and your indomitable spirit.

Once again, thank you for your service, your sacrifice, and your dedication. You are, and always will be, a beacon of courage and resilience.

CHAPTER 10:
RESOURCES AND APPENDIX

To help you continue your learning journey, we'll provide a list of resources, including books, articles, podcasts, and other sources of information and inspiration that we recommend for further reading.

Chapter 1: Laying the Groundwork - 3 Years Out
1.   **Military OneSource:** A comprehensive resource for all things military, offering a wide range of tools, tips, and services designed to help service members, veterans, and their families navigate life in and beyond the military.

Chapter 2: Building Your Brand - 2 Years Out
1.   **LinkedIn:** A networking platform for professionals across various industries. It's a great place to build your brand, connect with other professionals, and find job opportunities.

2.   **careeronestop's Military to Civilian Occupation Translator:** Helps translate military career codes into civilian occupations, aiding in the creation of resumes and understanding job descriptions.

Chapter 3: Strategizing Your Approach - 2 Years Out
1.   **O*NET OnLine Military Crosswalk Search:** Allows you to enter a military occupation code (MOC) to find civilian occupations that utilize similar skills.

Chapter 4: Gaining Momentum - 1 Year Out
1.   **USAJobs:** The federal government's official job site, which also includes resources specifically for veterans.

2.   **Glassdoor:** Offers a wealth of information on different companies, including salary data, company reviews, and job listings.

Chapter 5: The Final Push - 6 Months Out

1. **ClearanceJobs:** A job board and career network for professionals with federal government security clearance.

2. **Indeed:** A worldwide employment-related search engine for job listings.

3. **Veteran Job Mission:** A coalition of companies committed to hiring veterans.

Chapter 6: The Home Stretch - Your Last Month in the Military

1. **Military.com's Veteran Employment Center:** Provides job listings and career advice specifically for veterans.

This list includes additional suggestions that may be beneficial to you as you navigate your career transition.

1. **Veterans' Employment and Training Service (VETS):** A U.S. Department of Labor organization that prepares military veterans and their spouses for meaningful careers, provides employment resources and expertise, and protects their employment rights.

2. **Feds Hire Vets:** A U.S. Office of Personnel Management resource that provides information on federal employment for veterans.

3. **Hire Heroes USA:** A nonprofit organization that offers free job search assistance to U.S. military members, veterans, and military spouses, and helps companies connect with opportunities to hire them.

4. **Hiring Our Heroes:** A U.S. Chamber of Commerce Foundation initiative that helps veterans, transitioning service members, and military spouses find meaningful employment.

5. **G.I. Jobs:** A resource that offers job listings, career advice, education tips, and transition guidance for veterans.

6.   Veteran Mentor Network: A group on LinkedIn where veterans can connect with potential mentors who are also veterans.

7.   The Military Wallet: A personal finance and benefits website for military members, veterans, and their families.

8.   Office of Veterans Affairs; VA/DoD eBenefits: Provides up-to-date and comprehensive information about the benefits provided to veterans and their families.

APPENDIX I

Summary of Chapters

Chapter 1: Preparing for Transition Keywords: Military Transition, Career Transition, Civilian Job Market, Transition Timeline, Resume Building, Networking, Research.

Description: This chapter provides an overview of the military-to-civilian transition process and introduces the three-year plan for a successful career shift. It emphasizes the importance of early preparation, including crafting a compelling resume and building a strong professional network.

Chapter 2: Self-Assessment and Goal Setting Keywords: Self-Assessment, Skills Inventory, Career Goals, Values, Interests, Strengths, Weaknesses.

Description: In this chapter, readers learn how to assess their skills, experiences, and qualifications to determine their worth in the job market. It includes a worksheet to identify strengths, weaknesses, and areas for improvement, and encourages setting clear career goals aligned with personal values and interests.

Chapter 3: Researching Industries and Companies Keywords: Industry Research, Company Research, Job Market Trends, Job Search Strategy, Target Audience.

Description: Chapter 3 guides readers in researching various industries and companies to identify potential employers. It highlights the importance of understanding job market trends and tailoring job search strategies to specific industries and companies.

Chapter 4: Gaining Momentum - 1 Year Out Keywords: Active Job Search, Job Application Timeline,

Location Research, Salary Negotiation, Compensation Packages.

Description: This chapter focuses on the final year before transitioning, providing readers with a timeline for job applications and the importance of researching potential locations and salary ranges. It includes tips on salary negotiation and assessing compensation packages.

Chapter 5: The Final Push - 6 Months Out Keywords: Job Application, Interview Techniques, Job Offers, Career Transition, Skillbridge Programs.

Description: Chapter 5 delves into the critical six-month period leading up to the transition, offering guidance on effective job applications, interviewing techniques, evaluating job offers, and utilizing Skillbridge programs for skill development.

Chapter 6: The Home Stretch - Your Last Month in the Military Keywords: Emotional Aspects of Transition, Relocation, Housing, New Job Preparation.

Description: This chapter addresses the emotional challenges of transitioning and provides practical advice on final preparations, such as relocation, housing, and preparing for the first day in a new job.

Chapter 7: The Aftermath - Your First Year as a Civilian Keywords: Adapting to Civilian Work Culture, Performance, Networking, Promotions, Salary Negotiation.

Description: Chapter 7 explores the first year as a civilian employee, focusing on adapting to the civilian work culture, and achieving career growth through performance, networking, promotions, and salary negotiation.

Chapter 8: Career Growth and Long-Term Planning Keywords: Continuous Learning, Skill Development, Long-Term Career Goals, Professional Development.

Description: In this chapter, readers learn about continuous growth and long-term career planning, emphasizing the importance of continuous learning and professional development to achieve career goals.

Chapter 9: The Importance of Giving Back: Mentoring Others in Transition Keywords: Mentoring, Veterans Support, Giving Back, Transition Support, Career Guidance.

Description: Chapter 9 highlights the significance of veterans supporting fellow transitioning service members by providing mentorship and career guidance. It emphasizes the impact of giving back to the veteran community.

Chapter 10: The Lifelong Journey: Continuous Growth and Development Keywords: Lifelong Learning, Personal Growth, Professional Development, Career Success.

Description: The final chapter reinforces the idea of continuous growth and development as a lifelong journey for career success, encouraging readers to embrace a growth mindset and seize opportunities for personal and professional growth.

Appendix: Resources and Further Reading Keywords: Career Resources, Job Boards, Veteran Support Organizations, Books, Podcasts.

Description: The appendix provides a comprehensive list of resources, including career support organizations, job boards, books, and podcasts, to further aid transitioning service members in their career journeys.

APPENDIX II

Listed alphabetically by topic

ABOUT THE AUTHOR

Cheryl A. Cross is a recognized business consultant, workforce development council and executive board member, and advocate for military transition and military spouse employment. She is also the owner of C.A. Cross & Associates, LLC, in Hawaii, where she advises on Hawaii State-centered workforce development policy and creates cutting-edge corporate HR solutions. Her focus on helping companies effectively streamline or build recruiting and retention strategies, along with regulatory compliance within the Indo-Pacific region, positions her as a trusted advisor for companies navigating complex staffing challenges.

Her first book, "CIVILIAN MISSION: The 3-Year Guide for Military Professionals Planning Civilian Careers," addresses the urgent need for early preparation and self-exploration as military professionals seek to transition into civilian careers. The book reflects her commitment to equipping veterans with the strategies and confidence required to make this significant life change.

Aside from writing, Cheryl has leveraged her former radio broadcasting experience to develop a best-in-class media platform, XCHANGE. This podcast series platform will host multiple 12-episode series on empowering topics related to change and transformation, including the inaugural podcast, "The Military to Civilian Career Transition Power Hour," further supporting her mission to provide actionable advice to transitioning service members.

With an eye towards continued service to the veteran community, Cheryl is currently working on her next book, "Unlocking Your Worth: Salary Negotiation for Veterans," set to release in 2024. It focuses on empowering veterans to secure fair compensation in the civilian workforce. When not writing, working with corporate clients, or doing civic work, you can find Cheryl fishing from the beautiful shores of Hawaii with her trusted companion, JD (Just Dog).

www.ingramcontent.com/pod-product-compliance
Lightning Source LLC
Chambersburg PA
CBHW071558210326
41597CB00019B/3300